Ira C Fuller

The Mysteries of the Formation of the Earth

The Rising and Sinking of Continents

Ira C Fuller

The Mysteries of the Formation of the Earth
The Rising and Sinking of Continents

ISBN/EAN: 9783337404888

Printed in Europe, USA, Canada, Australia, Japan

Cover: Foto ©berggeist007 / pixelio.de

More available books at **www.hansebooks.com**

THE MYSTERIES

OF THE FORMATION OF THE

EARTH

THE RISING AND SINKING OF

CONTINENTS

THE INTRODUCTION OF

MAN

AND HIS DESTINY

REVEALED

In God's Own Way and Time.

———

AUTHOR'S EDITION.

———

BUFFALO:
CHARLES WELLS MOULTON,
1899.

INTRODUCTION.

"The contents of this volume were given by Spirit Josephine through the Mediumship of Mrs. M. T. Long-ley, the controlling Spirit Josephine—claiming at the time to be enrapport with a Band of Ancient Spirits from whom she received the matter which this volume contains. Josephine is a member of an advanced Spirit Band, who have selected the publisher of this work as the recipient of their intellectual and instructive favors. The various Intelligences of this Spiritual Band of workers have communicated to Mr. Fuller through different Media in such manner as could leave no doubt in his mind as to their veracity and intelligence, and as to the similarity of intellectual vigor and of personal characteristics as manifested through each Media."

CONTENTS.

vi   CONTENTS.

Tutelary Gods and Ancient Spirits.

IN THE seventh sphere or zone of spirit life, there is a grand temple of magnificent proportions and of sublime architecture. The material of which it is constructed is of dazzling whiteness and seemingly of some translucent stone, yet this substance had never been created by the potters' art, nor had it been hewn from any quarry of stone. The temple, so grand and lofty in design and finish, so beautiful in its every part, so dazzling in its whiteness, had not been erected by the slow laborious processes known to earth, but it has stood for thousands of years, a sparkling gem of architecture in the beautiful landscape, a tribute and a monument to the skill, the genius and the power of creative force in the human mind, for be it known to ye of earth, that this structure so extensive in its domain, so marvelous in its proportions, was brought into existance in the twinkling of an eye by the magical will and desire of Cabul, a God—yet a man––of the celestial world.

For once on a time Cabul conceived the idea of a temple that should be a poem in stone, a gem in architecture, a temple that would be vast and spacious enough to hold all the people of the country if they should care to enter its courts at once or to stand within its magnificent halls. And as he thought of the temple "not made with hands" eternal in the Heavens the picture of it, complete in all its parts, finished in every detail of perfection possessed his mind, and there upon he stood in the centre of a great and beautiful valley, surrounded by lofty mountains, whose crests were tinged at times with the ruby or golden glow of the east, or tinted with the purple shadows of the violet hued western sky—and as he stood, with concentrated thoughts and uplifted brow, he cried out with an intensity of fervor and command, "Be thou, oh temple sited upon this fertile plain, and be thou fixed as the mountains, and stable as the everlasting stars—Nor shall one stone be piled upon another, but thou shalt stand as one gigantic stone of dazzling purity that shall glow in the light even as pearl in its setting of priceless worth." And even as he spoke, there upreared as out of the empty air, a vision of a completed and gigantic structure, built upon a foundation of solid stone that glistened in the light—a temple with sculptured pillars and graven walls, a temple with its dome of splendor shining like an immense jewel in the light, and its doors were as of precious metal and frosted with delicate tracery of human skill—its aisles were broad and beautifully inlaid with mother of pearl, fountains played in its courts, and rich draperies adorned its walls, and the temple filled nearly the whole valley with its capacious proportions, pushing the flowers and the shrubbery to the very feet of the mountains that stood in majesty sublime as if keeping ward over the scene.

And all the people of the sphere rejoiced—for it had been told to them that one of their wise men should cause a temple to appear in a moment and that it should be known as the temple of Diernes, from whom the power had come to erect this mighty pile. Now be it known that Diernes is the tutelary Goddess that rules the seventh sphere—or rather that she and Diern as one being but as two entities, male and female are the tutelary Gods that rule the seventh zone.   That these great entities are not seen by men, but when they are ready to show special favor to the people of their zone, they select one who is of great will power to accomplish some marvelous feat, that will also prove a blessing to the entire race.

And thus had it been proven that the people of the seventh sphere were in great favor with Diern and Diernes for Cabul had been selected to produce this magnificent temple out of the air and by the simple effort of his will, that the people might have their own place of assembly and instruction in which the good of all should be shown.

Now there are many powerful beings like unto Cabul, in the wondrous realms of the seventh sphere, and we have only chosen his work as an example of the great and stupendous works that such as he can produce at will, and it is thus that we mention the temple that for thousands of years has stood in its pristine beauty and glory, unmarred by the lapse of time, and unstained by the approach of age.   For here is the realm of beauty and of light, there are no storms and wintry colds.   All is light, all is beauty and order, there is no sign of weakness and of age, for the glory of God descends upon mankind.

In the seventh sphere dwell a race of people who lived on the planet earth thousands of years ago—and beyond them from the seventh to the twelfth zone dwell people who were workers and thinkers in the mortal world, twenty and twenty-five thouands years ago.   These peo-

ple are happily endowed in culture, in intelligence, wisdom, and the power of achievement. Their works are of
such a wonderful character that they would surpass the
limits of human credulity on earth, should we attemp
to describe them. In appearance they are glorious—of
great height, some of them being ten and twelve feet
high, of massive proportions with faces that shine as the
sun. Their garments emit light as they move about, and
a trail of glory always follows in their wake for it is the
lumination of their being that streams out in beauty to
brighten all things in their path.

These are the Gods and Goddesses of the spheres—not
known as such by their followers or pupils, but simply considered as philosphers, teachers and guides, but corresponding in person and power to the idea of Gods and Goddesses which man on earth has in every age cultivated
or expressed. In the remote ages of the early human
history of life on- the planet Earth—beings of this class
who now occupy the zones of spirit life from the seventh
to the twelfth sphere as teachers and wonder workers—
occasionally visited Earth, and now and then some seer
or deep student into the occult sciences caught a vision
of one or more of them, or were let into the secrets of
the Heavens through a psychical revelation from some
one of these ancient guides--and then the glory of celestial life was told or sung to mortals, and it was affirmed
that God had come down among men to show the wonders of his power, and to give council to his chosen.
And the idea grew in the minds of the faithful that God
or the maker of the universe had a form like unto that of
man, only more glorious and majestic, and that He deigned

to commune with men on earth. Thus through legend and tradition, the belief in the personality of God grew, and He was mentioned as occupying a throne in the centre of the Heavens from which He commanded the earth and all that upon it dwelt.

But the Gods of the spheres are many and not one of them has assumed nor attained the height of Infinite power, yet their achievements are vast and wonderful, by the side of which the grandest invention and achievement that science has ever made through her most pow erful agents of earth seem as but the building of mud toys in the street by tiny urchins, as compared to the marvelous mechanics and constructors of the highest executive skill and art that mind and brawn on earth have ever shown.

The Zones of Spirit habitation from the seventh to the twelfth are not entirely peopled by those ancient intelligences of whom we write—on the contrary those Zones are principally inhabited by advanced human Spirits who dwelt on earth in comparatively modern times. These people make happy homes and follow grand pursuits in the spheres, and are constantly achieving more and more wondrous tasks, for they are under the guidance and instruction of the Ancient Beings mentioned before, who remain upon these particular Zones from choice, doing service to the dwellers therein, teaching them "the way, the truth and the life."

The ancients who dwell upon the lofty plains and pursue their beneficent works among the people who look up to them with reverence and veneration hold a special influence over certain portions of each sphere below them.

Thus Cabul as a wonder worker and a seer has an influence and charge over all the phases and labors of phenomenal mediumship on Earth. Not that he personally visits each medium and labors with him or her, but all the various intelligences who are employed from Sphere to Sphere belong to him in the work of producing physical phenomena through earthly mediumship and are under his control. Camaga, another ancient, is one who lived on earth twenty thousand years ago. He was a great seer, studied the stars and foretold events—he has charge over the labors and the manifestations of clairvoyance in spheres below him, including Earth. Darneva, another ancient, one who possessed grand healing powers, has charge over the healers and the physicians and they know nothing of his power; for the guides and familiar attendants of mediums do not divulge the personality of that source of energy or of influence, that comes to them through intermediary channels from one original Humanitarian fount and guide.

II.

We have mentioned the names of several of the powerful Ancients who keep watch and ward over the lives and missions of various people on earth and in the spheres below their own domain, and there are many such exalted characters, who come and go at will—nor are they limited to the planet Earth and its surrounding Spirit worlds, but they can journey far out into space, and hold communication with planetary Spirits of other planets and systems, bringing back the information they receive, and imparting it as they think wise and best to the human beings of the spiritual Zones who are prepared to receive it. In the seventh, eighth, ninth and tenth Spheres dwell many Ancient Greeks and Romans, as well as other people of various earthly languages and nationalities, Philosophers, Sages, Bards and Teachers, but while these have been in the higher realms for perhaps four or five thousand years, and although their powers of achievement are very grand, these personages are but as mere pigmies in learning and ability, and as infants in age and expression, compared to the more antique human beings of which we write, and in these papers we will not stop to deal with them—Cleopatra of the Nile is but a Belle compared to the ladies of high degree in spiritual realms, who have sojourned in their heavenly homes for twice ten thousand years.

Now be it understood by our readers on earth that these ancient beings did not all belong to one race or clime in the remote ages of the past when they dwelt on mortal shores, some of them lived in one part of the world, and flourished under one dynasty, and some, in another. Since their time on Earth the whole face of your planet—Oh, Mortals—has been changed where now are oceans and seas, lands and nations offered up their riches to the eye or grasp of man—where once, Oceans rolled in rythmic music and deepest harmonies, Continents now stretch from the shores, of one mighty body of water to another.

It is not an easy task to make you realize in just what locality the submerged continents of the past lie; but

there have been more than one country inundated, and
buried where the waters of Centuries have sung their
requiems for thousands and tens of thousands of
years. Even the most radical and enthusiastic of
scientists whose geological researches have convinced
them of the antiquity of Earth, have no idea how
old their planet really is—and were they to be told, that
fifty thousands years ago men and women lived on earth
—not as cave dwellers and uncouth savages, but as polished, cultivated, intellectual beings, who had a knowledge of the arts and sciences, they would scout the story
as that of some wild Munchausen, who was striving to
learn how far he could tax the credulity of the world by
his assertions and tales, and yet it is true that in the past
as remote as that mentioned in the preceding paragraph,
human beings of cultivated tastes and great knowledge
lived on earth and pursued their tasks and missions with
fidelity and skill, but as the centuries rolled by their
lands became inundated, their homes and temples destroyed
their possessions lost, their arts and sciences banished and
their races extinct, all because of the violence of Nature,
in her efforts to illiminate certain objectionable elements
from the planet, and to bring all people and all things
into a line of unfoldment that would ultimately produce
the highest state of harmony on Earth.

But, while these people were thus swept from the face
of the earth they were not annihilated, for their homes
were to be found in the spirit world. Such was their
advancement in all that goes to make up a truly Spiritual
selfhood—that when destruction came to the earthly,
they were at once prepared to enter the fifth and sixth

spheres, without having to be domiciled upon any of the Zones below that number.

By and by we shall tell something of the earthly Continents that these different peoples inhabited and also of their habits, customs, employments, and temples of those times, but now, we are dealing with them as Spirit entities above the Earth, and have something to say in regard to their work and life on high.

First let us remember that ere the planet earth was fashioned and sent rolling in its orbital line through space, that grand, majestic beings from older planets in the Heavens, who were hundreds of thousands of years old, had been commissioned and equipped with intelligence, wisdom and power, to operate upon the germ life of the planet and vitalize it with such elements as would enable it to grow and develope into planetary activity. Now, these tutelary beings—two in number, grand as Gods, male and female, had assigned to them as subordinates to do their bidding in the planetary work and wardenship, many human entities from the planets who are themselves as beautiful as the morning and as powerful in intelligence and expression as a mortal could conceive it possible to be, and it is such as these that are occasionally seen and known as planetary Spirits by media or clairvoyants upon the Earthly plane.

As we have said the earth was brought into working order under the influence of these high entities, and they still preside over its destinies, and labor for its development towards ultimate perfection. But these beings were not busy alone with the earth, they had something to do with its spirit Planet and Zones also, to magnetize them, or to infuse into them something of the human electrical and magnetic properties that would mingle with the forces sent off by the earth and its people by radiation and vibratory force, and help to bring them into proper condition for the habitation of Earth's children when they should throw off the body of matter and enter the spirit realm.

Let us now return in thought to the ancients who dwelt on your planet twenty or more thousand years ago.

They were wise and exalted. Why? Because certain of the Planetary Spirits perhaps a hundred or more, had years before descended from their estate, and under the care and counsel of the Tutelary Gods, had thrown off their own bodies, and as human souls without part or form had mingled with the essence of magnetic human life on Earth, being absorbed by the parental organs of human beings, and been born on earth of harmoniously mated beings, thus finding and bringing the conditions of Spiritual pefection and completeness into the mundane sphere, and so it happened that these planetary spirits became earthly beings for a time. They were endowed with grand proclivities and as they grew from infancy to youth, their pure and wise characteristics were so apparent, their intelligence so remarkable that they were given every opportunity to expand their powers and to cultivate their intellects, but the highest teachings and the grandest truths were imparted to them by the planetary Spirits who attended them, and as the years advanced, these gifted mortals became teachers and sages and high priests of learning and of art, and they taught the people of their lore. Most of them were quick to learn, being intuitive, innocent, and susceptible to high influences. Temples were built for the wise men and women, in which to found schools of philosphy, of art, of music, of literature, of science and of manual training. All the children were taken in charge by them, and for a period of twelve or more years kept under the direct teachings and guidance of the wise Ones. The result of all this was that a race of healthy, simple, pure minded, highly endowed and cul· tivated men and women was reared, and when they passed from earth they were prepared to find homes and labors in the spheres above the first five surrounding the earth, but the planetary spirits who had been elected for re-imbodiment, also passed from earth becoming at once what they were before, planetary Spirits, but only for a little time. Again they were reborn on earth, again did they grow to be teachers and wonder workers, and again lead the generations onward to grand achievements and power.

But at length the fateful hour came when the countries —each in turn, some of them many, many centuries earlier than others, were submerged and their people swept from the face of the earth, and again the waters subsided and new continents were born, these being peopled by voyagers from others countries and by the Spirit world— for remember that the whole globe has never been submerged at once.

We will now turn our attention to the Spirit Zones at the times when the first submergence of a continent occurred. These Zones were by that time seven in number, there was also a spirit planet. All of these were well vitalized by their spirit inhabitants and by emanations of magnetic human life sent off by the earth. But the Tutelary Gods and their helpers knew there was much work to be done. The Gods foresaw that strange races of men were yet to inhabit the earth, that art and science and philosphy would be lost to it for thousands of years, and that for a time it would descend to a state of savagery and darkness, ere it would develope a high civilization and evolve a more beautiful atmosphere and climatic condition.

And these Gods foresaw that the lower spheres of spirit life would be needed by human entities who would be cast from earth in a state of crudity and ignorance, and also that the first sphere would be darkened from its pristine light by the emanations from crude and sensual lives on earth and become unfit for the habitation of any but themselves, and that the succeeding two or three Zones would become stepping stones—or abiding places, for spirits in gaining experiences and in working towards the higher realms, and thus the Gods set the higher Spirits who had advanced from earth to exalted states to work in creating new Zones for the blessing of human souls.

III.

And the work of these exalted ones was to at once gather from the realms of space the floating nebulae of worlds and to breathe upon it and infuse it into the emanations from the lives of the inhabitants of the seventh Zone, which emanations encircled that Zone like a girdle of light—and as these magnetic works went on, and the new circle gathered density and strength under the manipulation of the exalted ones on the eighth world for spirit habitation grew into shape, and people from the seventh Zone moved upward to it, erecting temples and palaces, laying out parks and gardens, by the exercise of will power and magnetic vibratory force, and immediately the Gods and the higher Spirits of the eighth Zone went to work to create still another Zone which was accomplished in due time, and the work of ascension or removal went on, the inhabitants of each sphere ascending to the other, until the first, second and third were left for future use of earthly beings, nor did those who removed from one belt of existence to a higher have to occupy the dwelling or to accept the gardens of those who had preceded them, for the latter ere they took their departure—either by will force, or by electrical and magnetic impulse, caused all those to vanish that their successors might create their own dwellings and lay out their own grounds, according to their own taste and skill, for it is by such works and also by deeds of sacrifice that the human entity grows. All this building of Zones out of magnetic aura and will of human beings, supplemented with the planetary dust of the Heavens took time, and in the meanwhile the greater portion of the earth was passing through long periods of cold, desolation, and darkness. Its people had degenerated, only remnants of once powerful tribes existed, and these had to remain in the shades of discomfort and unhappiness, only a deteriorated class of once powerful races existed, and these were obliged to battle with the elements, nor could they regard the hither forces of intuition and spiritual light, but the germs of greatness did not die, the possibility of growth and the promise of future culture were there, and as the Earth trundled on, conditions were reaching it by which a high and lofty humanity should be reborn.

But the lowest sphere of spirit life was now being tenanted and its newcomers were people of feeble intelligence and of brutal instincts, undeveloped and ignorant human beings unto whom, missionary spirits from the higher realms came, as teachers and friends. Man on earth had been struggling with all sorts of adverse conditions, atmospheric disturbances discomforted him, need of the necessities of life made his position an unenviable one, but slowly through the ages, the natural energies of his being began to assert themselves, and he commenced to grow in intellectual vigor until his powers of calculation, of forethought and of judgment enabled him to plan and to contrive means and methods by which he might overcome the inclemency of the weather and to provide a means of subsistence for himself and family. And so through the centuries humanity struggled and grew, until it has reached its present status of civilization and progress, but never since the floods came and continents were submerged has mankind developed a race of such lofty, intellectual, scientific, and progressive people as existed in those prehistoric days—nor will the earth be populated by such an humanitarian and exalted people, until ten thousand years or more have rolled away. For in those days the races of earth were quickened and vitalized by the power of the Spirit to a greater degree than they have ever been since. Planetary Intelligences voluntarily taking up a birth and a life on earth for the purpose of educating and spiritualizing the people, brought to the earlier races of man an influence and a power from superior realms that lifted them above the ordinary levels of earth to a plane of intellectual and moral vigor, which the planet per se was unable to unfold. But, there is hope and there is promise for the future good of earth, humanity on the planet is constantly growing nearer and nearer to the Divine. The psychical and moral forces of human beings are operating all along the line of progress, and the race is rising age by age towards a grand and lofty altitude of mental and spiritual light—there is more progression and a steady moving onward and man, triumphant in his march realizes the growth of centuries and is glad. But at present we will deal with the ancients in their sphere of light on high.

From the eighth Zone that had been inhabited by advanced entities went out a powerful magnetic current charged with elements of stability and beauty which were vitalized by the electrical forces imparted to them by the Tutelary Gods and their planetary agents, and which in time created the ninth belt that swept in glory around the eighth, and when this beautiful region was ready for their habitation innumerable companies from the spheres below ascended to its realms, while the places they had vacated, were again filled by the promotion of the people in spheres below—and so on through the centuries new spheres being created one after another according to the need, and to the constantly unfolding intelligence of their creators, until a cycle of twelve Zones or belts surrounded the earth, each one peopled by sentient entities according to their Spiritual development and all completing the first portion of creative work set for the Ancients for their initiative labor in the creation of worlds, by their Tutelary and Planetary guides.

During all these centuries the Ancients busy with their own work felt and knew that there were still higher tasks in which they might eventually engage, for although by their energies and magnetic forces they could create a spirit Zone, yet they were not capable of building a planet or world in space such as are seen in the expanse of Heaven.

Tutelary Gods and their subordinates were the powers that performed such work, and although certain of the Ancients could at times catch a clairvoyant glimpse of them, the majority of these Zone builders had never beheld the powerful beings who led them on. However, the work of Zone building went on. After the twelfth came the thirteenth belt which was thrown out by and around its predecessor, and so on continually until twenty-four Zones or belts complete the Cycle, of the ages of the past, and as each Zone became completed, certain of the inhabitants of the sphere below it began to ascend to its regions of delight—but many of the Ancients remained on the various Zones, preparing to serve as teachers, guides, and wonder workers for the human family below them, rather than to ascend to higher spheres, and while many others did move on and on, in order to hand down their influences and instructions to their co-workers among the people of the lower planes, yet there are countless numbers of the Ancient philosophers, sages, and magicians, who still dwell upon the different Zones from the seventh to the tenth, whose works are as wonderful in the eyes of their pupils as would be the manifestations of Gods to the inhabitants of earth. The ancients who dwell among the people of any sphere below the thirteenth are not only sages, philosophers wonder-workers, artists and scientists in their own powers, but they are also seers and mediums and by their psychological qualities they are enabled to receive communications from the ascended souls beyond them. Thus they are doubly equipped in their labor of helpfulness to humanity. Every Zone has its Tutelary Gods—two, male and female—and these Tutelary Gods of the Zones are all under the guidance of the superior pair of Tutelary Gods who were instrumental in bringing the Earth, and its Spiritual planet, into existence.

For thousands of years the first belt or Zone surrounding the earth has been the abode of Earth bound spirits. Those human beings who have lost their mortal forms, but who have not developed sufficient Spirituality at their earthly decease to ascend into the truly spiritual realms, and who are therefore for a time obliged to remain in close contact with earth and to develope a higher faculty of perception before they can reach the light.

This first Sphere is veritably an intermediate place between Earth and Spirit life, it is the vestibule into the Spiritual realms, and to many souls it is a region of darkness and cold, because the emanations from such lives as are dark and turbulent and chilling—in their selfishness and iniquities — on Earth, make up its atmosphere. Humanitarian spirits from various spheres beyond descend into these lower states to work as missionaries, as teachers and as physicians among the lowly and hopeless ones there, and to guide them into a better path.

IV.

After a time the missionaries succeed in their labor, and the darkened souls are illuminated with an interior light, and quickened into activity, by the magnetism of their helpers, until they grow out of their unhappy state, and ascend to the second sphere in which they become workers, helping those below them whom they may reach with sympathy and love, and thus working out their own progress toward the higher realms.

Among the missionaries who labor with the denizens of lower spheres as teachers and guides are many of the Ancients, who take great satisfaction in imparting their wisdom and lore to such souls as are prepared to receive or to be guided by them.

The first or lower sphere of the Spirit life is a make up of strange classes, they are from every race and clime, and because of their propensity to physical life, they exhibit all the traits of character and habit which possessed them on Earth. All really selfish and exacting people, all licentious or brutal persons, all who are coarse, degraded and vicious gravitate to the first sphere—some of these may remain for many years in its darkened precincts and not even know that they have passed through death. Others soon throw off the chains that bind them, grow weary of the wickedness and sin, and under the influence of the missionary spirits, gradually evolve a higher selfhood, that enables them to rise to upper spheres. Some of the beings who inhabit the first sphere were of high position on earth, there they appeared to be better than they were. There are Kings there, and Queens who presided over the destiny of Nations. Priests and Bishops who held the authority of the church and the welfare of the people in their grasp, and various others of rank and name, but none are there, whether King or yeoman, priest or layman—who were unselfish and who tried to do right.

Many of those who dwell in the lower sphere have a strong hypnotic power, and when they come in contact with susceptible beings on earth they cause these sensitives to see what they desire. Thus, one who was once a monarch—but who because of his cruelty and selfishness did not develope a spiritual nature strong enough to take him above the earth—may appear to a sensitive as a grand exalted being clothed in purple and fine linen, while in reality in spirit he may be dwarfed and misshapen and be garbed in rags. One who was an autocratic and domineering priest on earth may appear to a sensitive as an Angel of light, while in reality he is of dark and forbidding countenance and form. But if the sensitives on earth cultivate their own higher faculties, and aspire upward to spiritual things they will not long be deceived by such pretenders, for the truth and the light will be made known to them. The Ancients have power over the lower forces and planes of life and therefore they are of great service in breaking up haunts of evil-doers upon the lowest plane, where corruption and plotting against the comfort or happiness of men abound.

All spirit entities who are above the lower planes emit

a light from their persons, the higher in the scale of advancement a spirit is, the brighter will he shine. Ancients who have dwelt on the higher realms for thousands of years appear to be all light—they shine as the sun but the glory of their presence dazzles the sight and confuses the brain of those who dwell in the darkness of earth.   Therefore these Ancients make use of agents in reaching the Earth bound souls, agents who are less brilliant and advanced than themselves, and these are sent out to prepare the way for the coming of an Ancient, whose light is a searching power that reveals the hidden things in life and brings them forth.   After the agent has succeeded in reaching and infusing some needed quality of magnetism into some infested place, where, perhaps hundreds of degraded beings congregate, one of the ancients will come forward and by the light which he sheds produce a rapid vibration in the atmosphere which affects the condition of the people, and stirs them up to a greater activity.   This reaches their encrusted spiritual nature and impulse, and awakens in them a desire to know something more of life, and to come to better things, and they are thus prepared for the teacher, who is to come and enlighten them.   Frequently hundreds of beings are thus stirred into new activity at once.

The Ancient is not seen by them as a human being, but only as a brilliant light, but the teachers and guides who are of the modern age, who take them in hand, when the Ancient has done his work, are seen and felt—some dimly, others more powerfully—and are recognized as beneficient human beings.

Just as soon as a dweller in the lowest sphere is awak-

ened to his true condition, and exhibits a desire to re-
form, and to grow out of his unhappy state, his own vibratory force is increased, and this enables him, to slough
off the murky. weighty elements, that hold him to the
earth, and to generate an aura that is bright and of a little better quality than that formerly made. This is magnetically strengthened by the elements imparted to him
by the teachers and guides. He then becomes sufficiently
freed from his thraldom, to cease to be an earth bound
spirit, and is assisted to rise to the second sphere. Here
he may be placed in some school, sanitarium, retreat, or
other place of protection and training, according to his
needs where he will be afforded the care he requires for
the encouragement and growth of his higher qualities.
There are no schools nor sanitariums on the first Zone of
spirit life, but there are places of restraint there where re-
fractory spirits are sometimes confined until they are prepared to receive the light, and the magnetism that will
enable them to rise to a higher state.

Every sphere of spirit life as far as we know, is peopled by human intelligences, although many of those
upon the lowest Zone might be mistaken rather for animal, than for spiritual beings, while those upon the highest planes, appear more as Gods, or beings of light than
as human entities. There are many other human spirits,
who remain for years in the homes and haunts of their
friends, or former associations on earth, and who know
nothing of the spirit worlds that rotate with the earth in
space.

We are told by celestial beings of antiquity, in the communications which they give to us through intermediary

Oracles, that in the centre of the Universe, is a realm of glory and of power, the vastness of which no human soul can conceive, and that in those infinite realms exist beings —who serve the highest Light and Intelligence of all being—and that these entities far surpass, all likeness to human Nature, so Divine and Omnipotent have they become, but that nevertheless they have evolved out of the human family age by age and in reality they are perfected human souls.

In contradistinction to this we are informed that upon the earth, interpenetrating its atmosphere and mingling with the aura of caves, and forests and mountain retreats, even, too, amid the haunts of men in city streets and in frequented as well as remote places, is a sphere of life that is neither animal nor human but which is between the two. In this sphere creatures dwell that are evolving conditions that will in time, bring them into the human kingdom, and enable them to become possessed of human forms, and born in such, of earthly parentage, such pigmies or elementals can only enter the human race and amalgamate with it, by forming a part of the lower savage tribes of humanity on earth, but when they pass out of that initiatory stage of human existence they will be vitalized with a spiritual spark that will send them on the road to Immortality as human entities although they may have to be reincarnated again and again on earth through one tribe and another to higher estates ere their mortal births shall end and their mundane progress be complete.

These elements have a certain degree of intelligence the life principle of them has existed before in one or another type of animal form. They are active, shrewd, sometimes docile sometimes mischievous. Sometimes men on earth become aware of their existence and make friends with or servants of them. Sometimes peculiarly organized sensitives are annoyed by them while others are amused by their antics and pranks. Many decarnated spirits even of an advanced and highly intelligent order know nothing of the existence of these pigmies, Elfs, and Gnomes, but the Ancients knew and understood the elementals and are instrumental in bringing them into a state in which they are fitted to be born as human beings, for it is the electric magnetic force of such Ancients that vivifies the life principles of the elemental into the quality of spirit activity.

V.

In the temples of the seventh spheres it is explained to us by the Ancient Guide who rules that Zone, how human beings are brought—as entities into the atmosphere of earth, and how they become possessed of a mortal form. Now there are several sources from which humanity springs, that is we mean human entities on earth, although primarily the life essence and soul germ springs from one Eternal and never changing Source.

In our lessons at the temple we are taught that the life element or soul germ exists as a point of light and that this springs from the Parent Fount of all Intelligence and Power in the centre of the universe. That these points of light endowed with the potency and possibility of In-

finite expression and unfoldment along lines of intelligence and power, are sent off by millions from the Central Light—and that they are borne by irresistable currents of attraction to various parts of the Universe, coming into the atmosphere of different planets, and absorbed
by such planeary atmosphere and influences as they happen to reach, which influence and forces impart to them
qualities peculiar to themselves. With the Soul germs
that are swept from the Central Light to the planet Mars
or that of Jupiter or Saturn, we have at present nothing
to do. These have their own mission, unfoldment and
character. So do tne life germs of the planet Earth.
Each Soul germ as a point of light in its original form is
surrounded by a mass of nebulous matter sublimated and
refined, that is its protective shield while it floats in the
atmosphere.   When it becomes absorbed by the magnetic
aura of some female on earth at the moment when the
human foetus is conceived, the germ feeds upon this
gelatenous element which supplies nutriment to it, until
gestation proceeds and the embryo can draw sufficient
nourishment and vitality from the parent stock. If it
can do this regularly and connected the embryo grows and
in time becomes a living child, for the soul germ grows in
strength and power as the growth proceeds, if the vitality
is not supplied to the embryo according to the processes
of nature the soul growth or expression is checked, there
is a premature birth and the soul germ is cast back into
the atmosphere to feed upon magnetic forces supplied to
it by the Ancients, until it can again be absorbed by some
human aura on earth and be born as a living child. Those
who enter the parental atmosphere as soul germs only, are

not as yet human entities, they are the undeveloped germs of human beings; if they are cast forth before they have had an expression and experience on earth—of at least a few months or years.

Which stimulates an activity in the cranial region and produces intelligence—they will have no conscious, sentient life, in the spirit, but they must be magnetically and electrically breathed upon and cared for by the Ancients who understand the law, and brought back into the environments of human beings on earth, to be born as living children. The embryo becomes active and its human faculties are quickened by stimulating forces from the spirit guardians, some little time before its mortal birth. At birth it is Started on the road to intelligent unfoldment, in some cases the development of its intellect is more rapid than in others—owing largely to the qualities and elements absorbed by it, either from its earthly parents or from its spirit attendants, or both. If such a child as this passes from the mortal form a few months after birth, it may or may not find intelligent growth and expression in the spirit world. Some of these infants gain sufficient spiritual and electrical vitality in a few months of such experience to enable them to take a hold of spirit life proper, others do not. But if they do not, they appear as enlarged points of light which are kept burning by the magnetism of wise spirits until they can take advantage of the law of reimbodiment and be born again.

While reading these statements the reader may naturally inquire if these are truths, why do many spiritualists claim to hold communion with spirits, who have grown from immaturity to manhood in the higher spheres, and who purport to be the children of immature birth who had no expression on the mortal side. We shall come to the consideration of that in this paper which is now dealing with manner and methods of human birth, and expression on earth.

The human beings who first come to earth as soul germs —or points of light—from the Great Central Power, are as a rule gentle, timid, somewhat quiet and of a peaceful character—although these qualities may be, and often are, more or less modified, increased or altered by parental influence, prenatal conditions, or planetary action under which they are gestated and born, after gaining an earthly experience of greater or less duration they pass into the spirit world and enter associations and environments there —after the lapse of centuries they have their choice, whether to be reincarnated on earth as mortals or not, or them there is no arbitrary law in the matter, they can decide for themselves. Some of these souls during their first journey on earth became philanthropic teachers and ministers of helpfulness to their fellow men, some are artists, musicians, or perhaps circumstances and training have made them lawyers, soldiers, warriors. It matters not, what they were on earth, some may have been perverted into criminals—they need not be reincarnated unless they elect to be. But the law exists by which they may take advantage if they wish. Perhaps one who loved his fellowmen on earth may himself have been a sufferer or weak in moral rectitude. After experience and teaching in the spirit world, his sympathies may have grown very broad. He beholds the suffering and the sorrow of earth, and longs to help appease the woe. He decides to return to earth—he acquaints some Ancient guide with his desire. The guide selects for him the pathway to be trod, and helps to prepare the conditions for the new birth. The candidate for reimbodiment is taken in charge by the Ancient, and is put into a magnetic slumber, during which his spirit body becomes very attenuated. When the time of transit comes the magnetized being is borne to the atmosphere of the expectent parents, and the spirit or soul entity of that one is drawn from its dissolving body like a stream of vaporous substance and is absorbed by the magnetic atmosphere of the mother. Here it remains in a quiessant state during the first three months of gestation, when it comes into contact with the embryo and in proper time begins to vitalize it until a living child is born under such conditions, and in spite of circumstances, as will make of him a leader in moral teaching and conduct, or a grand helper through sympathy and love of his fellow men.

This we give only as an illustration, others may be reborn under different earthly conditions and for different purposes. Each rebirth is an advance in point of intelligence, power, opportunity and labor, over the preceding one of the same entity. The spirit does not remember its former incarnations on earth or beyond, because its work and mission in the present has solely to do with earthly things, but in time, after it has passed into the supernal spheres it will see it all and understand every such reincarnated soul is under the guidance of an Ancient, perhaps two, besides being assisted by other more modern intelligences. Frequently these guides keep all knowl-

edge or remembrance of former re-embodiments from
their charge deeming it best to do so.   Every spirit entity
thus reborn on earth is a sensitive, the psychical power is
quickened, many of them are grand mediums, through
whom, some blessed humanitarian work is performed, or
some grand truths taught.

The spirit bodies of these entities have become so very
attenuated and etherial by the time they are yielded up
when the ego comes back to earth, that they are at once
dissolved into the atmosphere, their elements forming a
sort of aura around the gestatory form, as a nucleus of
the spirit body of the coming child.   Sometimes the first
expression on earth of the same germ of a human entity
is so feeble and lacking in vitality, that it has no special
claim upon its earthly parents—and sometimes the earthly
parents are so utterly dissimilar in every essential to the
offspring that there is no real kinship between them,
when the child under such circumstances passes out of
the body it has no parents.   Such a child is reborn on
earth of parents to whom it is spiritually attached and is
a kin, it may be he shall live many years and at length
pass on when he recognizes the parents who preceded him,
or he may pass away first, and welcome his parents when
they come.   On the other hand, although now a vital entity because of his former embodiment, but now stamped
with certain qualities lacking before.   Qualities caught
from the new — and — real — parents and from various
sources, he may not be able to hold on to the mortal, and
may be expelled or sent back into the spirit realm. There
as an immature birth.   Yet the spirit claims the parents
he cannot reach in mortal, as his own, and it is such chil-

dren who manifest through mediums, as spirit entities, and claim kinship with the parents to whom in soul power they were drawn.

VI.

There is no inconsistency in the statement often made that children who had not matured on earth, can live and grow in spirit and communicate with those who would have been their mortal parents, had they been properly born and reared. Nor, in that of other spirits having to be reborn, in order to gain a vital hold of immortal life. All this we have endeavored to show in the preceding papers. It is rather unfortunate that we have to depend on mortal forms of expression and phraseology in depicting these things, because such forms are inadequate to our needs in this work.

The law of reimbodiment exists, it exists for the needs of any soul that desires to take advantage of it. Many human beings who are idiotic on earth are but undeveloped soul germs, for them the law is beautiful, and adapted to their needs, others who were bright and brainy on earth, but who desire further discipline, and experience on the mortal plane find the law a beneficient one to personal needs.   In cases where the soul is sufficiently vitalized, to pursue its growth along lines of intellectual unfoldment, the law is not an arbitrary one, for every intelligent being may decide for himself whether he needs or desires to be reimbodied or not.

In cases where the human entity is deprived of stimulous for its growth, or where the soul germ has taken no hold of immortality as a personal entity, the law is an arbitrary and a beneficial one.

There are other cases in which this law is a power for good although not operating as one of reincarnation in human form with these beings until later on in their development, and in our efforts to explain, how humanity through its various forms of creation—if we may be permitted to use the word—comes into sentient and intelligent being on earth, we cannot ignore this branch of our theme.

In the course of this narrative of Tutelary Gods and Ancient Spirits, we have mentioned elementals, who exist as strange and diminutive creatures, that appear to be neither human or animal in form and intelligence but a grade of being between these two states of active and concious existence.

From the knowledge we have derived from our Ancient teachers we learn that these elementals are formed from the essence, life principle, and sublimated elements, of certain forms of animal life which have evolved a consciousness and activity above the purely animal, yet not entirely of the human. That these creatures are real existences, more of matter than of spirit, and only invisible to mortal sight and touch because of the rapidity of their vibrations which the slower, duller senses of physical consciousness cannot reach, but that these same elementals often have a mischievous influence over certain human beings. The explanation is made that these elementals, are yet to be vitalized into spirituality by the soul germs of intelligence that spring from the great fount of Being, and that each of these creatures are capable of absorbing

one of these soul germs, and thus coming into contac-
with sentiment human life, of both the spiritual and the
physical plane, it becomes amalgamated with the higher
essentials of conscious activity. Attracting to its self a
soul germ, an elemental is in turn drawn with irresistable
force into the atmosphere of a human female in the lower
ranks and types of humanity—such for instance as the
Australian bushmen and it attaches itself to such a person.
In time, the female forms the necessary connections and
relationship with her male partner to conceive and bring
forth a child, that child is an embodied elemental vitalized
with a soul germ, but the lowest type of human life has
not sufficient spiritual vitality to make its way and develop
into active spiritual consciousness in the spirit spheres.
The child grows on earth and lives its allotted time, it
may be few or many years. It dies, the life principle does
not go back into the elementals, nor does it ascend to the
spirit realms, it remains in the atmosphere of earth, resembling a blue flame without parts or consciousness. In
time it is absorbed by the magnetic aura of some fertile
human female of a higher type of humanity than are the
lower forms of human life, and it is reborn as a living
child, now vitalized with sufficient soul energy and spiritual impulse to start it on its upward path of progress as
an immortal entity, in this case the law of reincarnation is
a necessity, an arbitrary and a beneficial one.

It is the Ancient spirits who lived on earth, many thousands of years ago, who have dealings with the law of reincarnation, and who understand it sufficiently, to give
wise expositions of it to mortals, did they feel it necessary
to divulge their knowledge of it. But there are few of

these very ancient people who hold direct communication with denizens of earth, although many of them have charge of sensitives and media, upon this and the other planes of sentient being, now and then intelligences who dwelt in the realms of the past on earth, in periods so remote that the annals of the time of Moses and the patriarch are pages of modern history, as compared to their records, they do find mediums in this age of the planets unfoldment whom they personally direct or control for some special purpose of achievement or object of guidance, but the instances are rare, for as a rule the Ancients labor through intermediary agents or messengers and pursue their way, unknown to or by the people of earth. We have given a slight explanation, of the birth of souls on earth, and the subject of pre-existence arises. That the soul germ starts out somewhere and some time on its pilgrimage and growth as Light we are assured, although previous to its appearance on earth as an intelligent and conscious entity it may have passed through many experiences, or previous or subsequent—to that mortal birth, it may be the creature of many incarnations. When the soul is thrown off from the great central soul and Sun of all Light, it is taken under the care and inspection of Tutelary Gods—or planetary spirits who build worlds—and it is never lost sight of during its eternal growth. Starting out as a point of light, it takes up its round of experiences and of existences while millions of years roll in silent grandeur through the ages. But as these cycles of time and of experience revolve, the soul is gradually growing toward the great central Sun of all infinite wis-dom and Intelligence—that which man calls God, and as

it thus grows onward it loses not one grain of its own individual consciousness.   But forever on it sweeps, gradually
becoming a God of infinite light and power, a Being of
Light forever more.

Finite man on earth asserts that as the beginning of a
straight line must have an end, so man if created, or sent
off from a greater force, having a beginning as a human
entity must have an end, but we assert that the premises
and conclusions only deal with physical and finite things,
that man starts out as a point of light, containing within
its self, the potency and power of self existence and of self
unfoldment.   That this potency and power is fed and kept
alive—no matter through what darkness or vicissitudes
the ego may pass in its disciplines by the electrical force
it draws from the great Central Light, and by the magnetism exerted upon it by the Ancients or the Tutelary
Beings, until it has developed its own inherent self sustaining forces that like the motion of the planets is per-
petual in activity and power.

The human soul starting out, as a point of light, gradually developes into a perfect sphere of Light—it glows
and scintellates within the human form, sending out a
halo that is now and then only perceived by keen sighted
eyes of earth.   Its radiations extend through the solar
plexus, and is distributed by the system of nerves to
every part and portion of the human organism—with
highly developed entities, these radiations not only illume
the entire being, but send out gleams of light that brighten
all things in their path.   These radiations that stream
from the soul centre, to and through all parts of the
organism, creates an active vibration, that according to

the spiritual statue and growth of the individual, produce emanations and auras of magnetic force which clothe—so to speak—the light rays and forces of soul, when the hour of physical dissolution arrives, the soul force ceases to vibrate so rapidly, it concentrates its powers upon the withdrawal of its light, and its magnetic aura from the body, and it begins the ascension from the head, gradually drawing to its self all the forces and elements belonging to its life, which go to make up the covering, or body spiritual, through which it is henceforth to act and serve.

The aura, or emanations of an active soul force, not only infil the organism and clothe the inner being, but they also send out forces and elements that cluster in a sphere around the individual as a protective shield of magnetic life—which in turn can be imparted to the sick and depressed as a life-giving force—the active soul generating the aura, according to its needs and works.

VII.

We have told of the birth of souls on earth, and of their unfoldment in various degrees—through possible successive Incarnations and in other ways. We will in these pages tell something of the rebirth of souls as spiritual entities, from the dissolving mortal body. But first we must consider, what is spirit and what its relation to and connection with the human ego. Let no man mistake Spirit as only a part of himself, something possessed, but rather let him consider Spirit as the Ego Himself or the entire humanized and individualized Being. We are told that "God is Spirit—and they that worship him, must worship Him in Spirit and in truth," and in considering the Infinite Light and Power, as perfected Intelligence as organized Wisdom, manifested through its body which is the Universe *per se*, we can realize that God is Spirit and as such the Eternal quality of Conscious Selfhood must only be Adored.

Man in infinite and limited conditions cannot behold the entire Universe of worlds, nor note the Infinite fields of space, therefore he can have no adequate conception of its form and parts—in their completeness. By the Ancient teachers we are taught, that man is a microcosm of the universe. We are also taught, that the universal ether is the breath of God—or Infinite Intelligence—and that the universe itself is the body of the Infinite Intelligence, the Soul force of which is the Central Light in the Heart of the Universe. As the Soul of Deity can only be conceived as Light so the Soul of man is Light, the life principle of Being, and as God or Soul Force in intelligence and light, radiates and breathes through the entire Universe which is self created and organized by this Infinite Soul Life, becoming an individualized Being.

So man is by the Soul force created an organized personified entity. God, the Infinite, centered in Soul force in the middle of the Universe is eternal Spirit by being co-ordinated as light, force and energy with the elements that it has thrown off and organized as a living form.

Thus too, man as primarily Soul Force, throwing off and creating forces, and elements that it organizes into shape and substance, becoming through his co-ordination

with these elements a living Spirit, Spirit being the result of Soul action and becoming the vehicle through
which intelligence is displayed as individualized and conscious Ego. This in turn creates activity in the molecules
and elements of matter that are energized by the electrical
force and magnetic impulse of Spirit, and organizes the
temporal body which clothes the indwelling Self—selfhood
that is the real personality, and which enables Spirit to
exercise its proclivities in a mortal career.

Soul force and Spirit combined make up the individualized eternal Selfhood that forever lives in conscious activity and Intelligent Knowledge and power, and as the
Spirit entity. As the Soul force of Infinite Deity as a central power, is located in the centre of its body—the Universe—from which it radiates its energy, to every portion
of its stupendous frame—so the Soul force of a human
Ego, is centred in the solar plexus of the individual form,
whence it radiates its energy and light to every portion of
its Being. The soul force is not situated in the Brain,
although the latter holds a concentration of force owing
to its delicate construction and machinery—which however is no more wonderful in its processes and functions
of utilizing and expressing the intelligent energy sent to
it by the Central Light, than is the cardiac organ wonderful in its processes and functions of activity in connection
with the circulatory life of the body. Intelligence can be
conveyed through any ganglionic plexus of the body—so
even can it be expressed by the finger tips of sensitive individuals, and the gray matter of the brain is not peculiar
to the cranial structure, for traces of it can be found in
various portions of the human form.

A man may break his neck and life departs from the mortal form, so too may one be shot through the heart or in some other vital part of the body and the spirit be expelled from contact with that external form. But the seat of life—or the Soul force of the individual is not at the base of the brain, even though in breaking his neck he is cut off from earth, nor is it in the cardiac region, but in reality it is situated nearer to the centre of the body, even in the solar plexus itself.

The Soul is light—the solar plexus is the lighted network that gives activity to the human frame. The organs of generation are also situated in the middle region of the body, and they are—in the highest conditions of functional activity—controlled quite as much by the Soul force as by the physical impulse of the Ego, Self. It was not ignorance, nor was it lust that prompted the phalic worship of the Ancients on earth, but in their reverence for and worship of the organs and processes of generation, they beheld the workings of infinite Life and Power, and recognized the all pervading glory and fructibility of creative soul energy, force and light.

And now having assured our readers, that the Soul is the vital flame or light that vivifies all things, and that Spirit is the correlation of forces and elements that Soul force attracts to its self, and which are vitalized by the Soul flame or life principle, the whole making up an intelligent entity, we will proceed to a consideration of the processes by which a spirit entity is born into the other life from the earth form.

As we have said at the hour of mortal dissolution, the Soul energy gathers to itself the auraific principles and energies which it needs and with these it rises from the body emerging from the head, and gathering around the body in a cloud-like substance, it seems to float at first, like a shadow in the atmosphere. In the centre of this cloudy vapor may be seen a point or star of brilliant light —which is the Soul or vital flame, and from this radiate rays of light to every part of the glorious cloud, gradually the (waves) of vapor assume shape and human proportions, the head and upper parts appearing first, from its filmy drapery, the lower limbs and feet appearing last. The brilliant centre still pulsates and glows, and can be discerned by the true clairvoyant eye. From this lighted point a slender cord of light is seen stretching from the newly formed being, to the fast becoming inanimate clay that was once called a man. This is the magnetic cord that has united spirit and body, during all career of the latter and it corresponds to the umbilical cord that unites mother and child during the gestative period. But now the spirit body is formed, the soul no longer vitalizes the mortal frame. The magnetic cord becomes loosened from the stiffening clay and the processes of material death are completed.

Nature is now satisfied to take care of the mortal through her agents of dissolution and change, while she sends the triumphant spirit upon its lofty march through the spheres.

We are writing now of the ascension of a good man— that of a viscious one would not be so beautiful and high, his spirit body would be composed of dark and noxious elements, and the soul force could not glow and palpitate

in light through its density, although the processes of
death in his case would be the same as in that of the unselfish man. We become the clairvoyant eye and behold
the ascension of the pure man. The aura arising from the
departing spirit, is like a billow of foamy lacework as
beautiful as is the glory of the dawn, it circles around and
envelopes him, and as we gaze in interested awe we dis-
cover that it shapens into curves and folds around his
form until it flows around him in graceful garments that
become him well. And now he is attired and ready for
his upward march, his face beaming with the inward light
is turned towards the stars. His form palpitates with new
life, he is equipped for the new life, and he is not alone.
Beautiful beings wait upon him and attend his way.
These are former companions, friends and guides while
they are watched over by Ancient beings, who have been
superintending and assisting in the closing earth scene
which separates a spirit from the mortal form.

And now the magnetic cord is severed, the earth form
is "dead" the spirit form is alive, it cleaves the upper air,
passing the first, second and third Zone, it takes no cog-
nizance of them—its attractive force, its affection and
affinities are beyond. Perhaps its speed is slackened at
the fourth sphere. If its home is there, yea, it involun-
tarily decreases its speed and pauses—its Zone is reached.
The spirit floats onward again, he has entered the fourth
world of spirits, he gazes around him at the enchanted
scene. Mountains, valleys and rivers meet his view.
Gardens, lawns, and flower bedecked roadsides invite the
eye, pretty homes and stately dwellings are before him.

Kindly faces and outstretched hands welcome him. He
is at home.   Old scenes are swallowed up even in memory
for a time by the beauty opening before him.   He is at
Home, a beautiful house stands open, it is filled with
comfort and delight—he enters it and is made welcome.
Tender ministrations prove to him the deathlessness and
the glory of human love. He is at home. The old time
worriments and sacrifices lose their sting, the shadows
have given way to sunshine, the winter of life has been
turned to golden Spring, a smile of content is on his lips
and in his heart, he is at home.

VIII.

In relating in substance what we have learned from our
Ancient guides concerning the birth of the Soul, its experiences through matter, its expansion and growth,
through one or various earthly incarnations, and its advance as an intelligent Spirit entity into worlds beyond,
we have shown that the Ancient intellects of by-gone ages
are thoroughly alive and active, and that they are occupied in the practical affairs of human life upon this planet
as well as in the spirit sphere. Individuals who wonder
what humanity can find to engage its attention during the
ages of eternity, have no conception what ever of the vastness of Life, its scope and possibility, nor can they grasp
the idea of Infinite Unfoldment and achievement.   To
Spirits who are busy building worlds, guarding and guid-
ing planets in their evolutionary growth, superintending
their affairs and progress and destinies of races of men,

and of Nations, supervising the birth and the expnsnaoi of Soul germs, looking after the fate of elementals, gui-d ing and guarding re-embodied beings through their one or several stages of mortal experiences, studying the Heavens and the methods of life upon the various planets, experimenting with the electrical and magnetic forces and currents of the Universe, creating wonderful productions of Art, Science, and of magical skill, studying the Occult and psychical laws and conditions of being, and achieving glorious results from all this reflection, research, effort, and expenditure of force and energy. Time is nothing, they take no heed of its passage, a thousand years is as but a day, and the birth and passing of a nation or a race from earth is as interesting and important to them—and no more—as is the development, purpose and results of a colony of ants, or a swarm of insects to the scientific mind who studies their habits, customs, and nature, and who trains—or guards—these creatures at will.

To such who have seen Nations and Continents rise and disappear, who have passed thousands of years in the world of Causes, dealing with the laws of Nature as the skillful mechanic deals with his tools, for the purpose of producing finished works with them, Eternity is none too long in which to study, to plan, to labor, and to achieve.

When we started out with these articles, it was our desire and intention to describe "The wonders of the Heavens and the glory thereof," but as we realize the lim- itations of phraseology and the boundaries of comparison and of similitude we despair in our task in giving an ade- quate conception to mortals of the grandeur of that life, which the beings who figured upon the planet Earth twenty-five thousand years ago had in the celestial sphere. We will, however, at this juncture select one of these Ancient characters, and tell you something of the scope of his influence and the grandeur of his works. This intelligence is known by the name of Iberna—a name that is significant in its meaning, and which literally interpreted would read—"Light bringer."—This Being when seen by the finite eye appears as a majestic personage of massive form and of benevolent countenance—and, but for the peculiar light that radiates from face and figure, he might not be taken for other than for any one of the Intelligences of Antiquity who might have sojourned on earth from two to five thousand years ago.

But the Earthly history of his being was far back in Antiquity, his home having been on the Ancient continent of old Atlantis where he grew in stateliness of character, and exercised his office of high priest, and teacher in the Temple, made sacred to his people by the light and influence which infilled it from above. To be a high priest in his age did not mean to be a functionary of the church, nor a proselyter of creeds—but it meant to be a holy man, a devotee and mystic. To be a spiritual guide, also a teacher and a physician. Hence, as high priest ministered to the mental, Spiritual and physical needs of his people—he was a father beloved by his children, a teacher followed by his pupils into the various pathways of knowledge which he opened to them, a physician studying the physical comfort of his patients and showing them by precept and example how to keep pure minds in clean and healthy bodies, and a spiritual guide, whose ethical and psychical inculcations and revelations, appealed to the intuition and the reason of his flock.

Such was Iberna, the high priest of the Temple Aruna, on the continent of old Atlantis, in the city of Humenia, a crescent shaped city, backed by the glorious mountains of Prunia, whose internal fires once in a thousand years belched forth their molten offering to the sea; and fronted by the beautiful blue waves of an inland bay that flowed in rythmic music at its feet.

In the city of Humenia, beautiful buildings of tinted stone, and other grand structures of marble for private or public assembly reared their spacious walls, and groves of fruitful trees, the foliage of which glinted in the soft and mellow sunlight like jewels of rare and brilliant beauty. In the heart of this crescent shaped city—from which, on either side stretched the right and left horn of the crescent until they tapered into the sea—stood the famous temple Aruna, in which the glory of the heavens, and the mastership of man over the kingdoms of life below him, were revealed to the studious mind. This temple was built of solid stone that gleamed and glowed like a great pearl, as the sun arose over its turrets, or sank to rest amid the waters of the sea. It had been built by intelligent master workmen, each of which had faithfully performed his part in its erection, from the hewers in the mountain quarry who brought forth the shining stone, to the master mason who superintended the whole, who in this instance was the high Priest himself. A temple of those days was built upon scientific rules in reference to geometrical lines and astronomical relations, and in itself stood as an unsealed book of information and study to the minds who cared to study its parts and whole. Each workman was himself a student of the universe, and in his part of the labor made special study of its relationship to the whole, that when his task was done he might be fitted to enter a higher class and pursue a more intricate branch of research.

Within the walls of this temple were various halls, each finished for and adapted to its use, in which the sciences or the philosophies of life might be studied and investigated and while there were numerous servants and teachers engaged in the instruction of the people, Iberna, the High Priest, was the guardian and principal over all.

We have related in preceding pages that in the early history of humanity certain planetary spirits descended to earth under the care of the Tutelary Gods who control this planet—and became embodied as earthly human beings, the High Priest Iberna was one of these. Born of humble but intelligent mortal parents he at a very early age manifested signs of great intelligence and power, and as he advanced in years he gained a magnetic sway over the people—exhibiting also such evidences of superior culture, intellect and wisdom—that he by acclaim became the acknowledged head and teacher in that dominion.

The habits of the people were those of modesty and simplicity, their garments consisted of robes flowing in graceful curves and folds from the shoulders, girdled at the waist by belts of finely wrought material. The females and males dressed mostly alike, though the former assumed a sort of mantle or cape which they wore over the shoulders, and their sandals were of a different pattern and texture than those of the men. The fare of the people was frugal and simple consisting of fruits and nuts also of bread from a certain kind of ground corn that was very sweet and nutricious.   They ate no meat, and there was no sign of animal life whatever, in the city of Humenia, although in the mountain districts, and in certain forest places of Atlantis forms of animal life did exist.

The people of Humenia as a rule were gentle, law-abiding, progressive citizens—they lived in refinement, and they dressed with neatness and care.   They were taught that Spirit and body equally required culture—that the body was the temple of the Spirit, and that to keep the body pure and clean and tastefully dressed was to pay proper respect to the spirit.

Their religion was one of service to mankind, and the living of a pure life as a preparation for the beyond. They believed in immortality, and that those who died ascended to beautiful gardens that hung in space, where they assumed charge over their people below them, and in which they fed the flowers and fruitage from the magnetic emanations of their own lives, thus keeping them in perennial bloom, while for their own nutriment they absorbed and inhaled the aroma of fruits and flowers and were refreshed.   They also believed that the dead at once begin to grow into a state of youth and beauty, from which they would never depart.   They paid tribute to the sun, as the creative force of life in Nature's works and also as the type of the highest Soul principle of which they could conceive.   They were not idolaters—but they held reverence for all that was true and good in universal life.

IX.

The people of these olden times, did not die young, there were no immature births. Living a natural Life, they remained on earth the full length of Nature's term, each generation growing wiser than the last. In the Temple, Iberna and his associate teachers frequently received instructions and revelations from the higher invisible realms, which in turn was imparted to their charges according to their needs.

The modes of locomotion or transportation in those days were by mechanically worked carriages and air ships, and by vessels upon the sea. The electrical and magnetic forces and currents of the atmosphere and of human nature, were understood and applied to practical ends.

The secrets of Art and Science were known and utilized. Silken fabrics were worn from the cocoon, and cotton and linen spun from the products of plant life, while beautiful dies of gorgeous or delicate hues were extracted from the products of field and wood according to the highest skill.

And thus Iberna ruled with gentle guidance and polished intelligence until many, many years had numbered their flight above his head. But at length there had been a strange element introduced into his kingdom. Travellers from afar had come by ship into the bay, bringing with them strange ideas and customs. Where they came from they did not tell, their language was not of the

Atlantians but they told of other seas, and continents that existed far, far beyond the Atlantian coast. Some of these people remained but a little time and then departed, others elected to remain at Homenia, and to cast their lots with the Atlantians by whom they were made welcome. And the years went on, until it became apparent that the strange ideas of the new settlers were having an unpleasant effect upon the Atlantians. Iberna called his people and admonished them to prepare for the coming of a great catastrophe, for it had been revealed to them that the mountains were soon to belch forth their fire upon the Continent, and that the waters of the sea were to sweep in, tearing down the safeguards of the bay, and surcharging the land with flood. At first his people listened with dismay, but ere long they forgot his harangues, for the foreign element that had mingled its life with theirs, scoffed at it. Again he warned and extorted, bidding them to prepare to leave their homes, to build them ships by which they might escape the fury of the storm. But the mountains reared their heads in silent majesty above them, the sea slept serenely at their feet. They could not believe that danger threatened, and they disobeyed the command of the master, who would not desert them. And thus they lived until the hour of doom. It was midnight when the storm arose. The wind howling in fury, the waters of the bay leaping up like mad things as in search of prey—the mountains flashing fire, as if the heavens were rent in twain—this was what met the terrified senses of the people who fled from their dwellings, which rocked around and beneath them. The earth trembled, they knew not where to flee. Hope and succor and rescue denied them on every hand. Suddenly in the glare of the lightning they beheld the form of Iberna arrayed in spotless white standing above them upon the dome of the temple, his hands raised in prayer, his face shining with the light that never was on land or sea—and above the fury of the storm his voice in supplication and blessing was borne to their affrighted ears.

But even as the vision came it vanished, for intense darkness covered the earth, in which no man could see the face of another and only the moaning winds sang a requiem to the doomed and lost. Then, a discharge as of ten thousand batteries of artillery, a shaking of the foundations of the earth, the belching forth of mountains of fire and smoke, the washings of a mighty tidal wave—the mingling of human shrieks with Nature's cannonade. Then silence and oblivion, for a continent and a Nation had been destroyed. But—no, all vestige of a World and of a people had not disappeared at once, for he, the High Priest, was still alive, just as the darkness settled over him, Iberna descended from his station into the basement of the Temple, and when the ruin came he was still unhurt, although surrounded by falling boulders and columns that hemmed him in. He felt the earth shaking under him, but knowing that his time had nearly come, he had no wish to escape, even if it were possible for him to do so. He only grieved for his people that none should be left to tell the tale of the last days of their Nation, although years before when the strangers had come to them, who had again departed, a few of the young men of Atlantis, had elected to go with them, and it was surmised that they had found a home, somewhere on the planet where they might engraft their customs upon and give their legends to another race.

By and bye the tempest subsided, the waters lost their fury, the lava burned in sullen blackness unto death. Days passed and Iberna was still alive, and now he managed to creep from his dungeon and to look abroad upon the sullen restless sea that swept to the very base of the Temple on either side revealing no trace of garden or home, only in the distance a solitary mountain peak. The Temple had been built upon a high eminence in the centre of, but in the back of the crescent shaped city, and thus it was that while all else had been submerged the basement of the structure was just above the face of the sea. But desolation, desolation reigned on every hand. Iberna gathering his robes around him, stood forth and in tones of lamentation mourned for the people who had turned the deaf ear at last to his counsels and who were now no more. His voice betrayed to the listening wind the sorrow of his heart, but only desolation stared and mocked at him and drawing his robe over his face, he turned again to the silence of the temple ruins. Thus passed hours in which the Ancient communed in silence with the planetary spirits of the air—and again he stepped forth. His face shone as the sun at midday when the atmosphere is clear.

The storm had again arisen, and the waters lashed the ruins with tempestuous force. He was alone, alone with the grandeur and the fury of an awful desolation but he quailed not. Stretching forth his hand he cried out in exultant tones, as the waters rushed in upon him, sweeping on in a restless flood, submerging all things in their path, and bearing the form of the old Patriarch down, down to a watery fathomless grave.

To the clairvoyant eye of earth that beholds him, the Spirit appears as the sage and teacher, of massive form and benevolent feature as he did to the Atlantians in ancient days. Thus does he also appear to many of the Spirits of the spheres who know of his work, but to certain exalted intelligences, he appears as a brilliant sunlighted Being whose rays stream forth in glory on every side.

He has not always occupied the high position that is his, through the centuries he has been climbing from sphere to sphere. Iberna in conjunction with his female counterpart, and with a train of kindred planetary Spirits, has charge over certain portions and missions of earth. His work is with sensitives who are fitted to be co-workers with ascended Spirit Intelligences in good works. A large part of his mission is to oversee and direct the work of his pupils in uplifting, training, and educating earth bound spirits, also those who are in the second sphere. Many sensitives on earth are observed by departed beings —with these. Iberna and his co-workers labor, to free them from thraldom, and to bless both the obsesed and the obsesing ones.

He and his army of intelligent spirits of different spheres including mortals—also are busy in bearing magnetic healing forces to the sick of body and mind, and they are busy in stimulating the forces of many sensitives, whose psychical powers are unfolded under their care.

The Ancient herein mentioned was not the only high Priest of the Continent Atlantis—there were many such throughout the land, but he was the presiding genius of the City of Humenia and its surrounding districts, and we have chosen him as a type of his class, that our readers may learn something of their powers.

Iberna wields an influence upon the Spirit Planet belonging to Earth, and over the seven surrounding spheres. He and his coadjutors are in communion with still farther advanced spirits, and they in turn are related to the Tutelary Gods.

Such an Intelligence as this Ancient Sage, does not need a special dwelling place or structure—the universe is his home and the elements of the air his servants, though he has temples and dwellings of beautiful design, and of grand proportions at his command, in whatsoever region he cares to move.

X.

Such Ancients as Iberna, Cabul, and others that we have mentioned with many other intelligences, who have passed ages in the Spiritual worlds, are fitted to preside over continent and Nations. Yea over spheres or Zones of immense breadth and extent. They are also able to travel from planet to planet of your Solar system and to communicate with the human spirits of advanced wisdom and power who dwell upon the spirit worlds of each.

In the seventh world of spirit that surrounds the planet earth, is the magnificent temple mentioned in the opening pages of this writing, as the instantaneous work of Cabul, the Ancient Sage and wonder worker.   Within the cen-

tral court of this Temple is a glorious shrine erected to
Truth, a shrine that gleams with the splendor of the un-
dimmed sunlight, and which sparkles with the beauty of
many gems.   At this shrine the wise men and women of
the seventh sphere, who are in communion with the celestial beings of worlds beyond them, receive the revelations
that such beings make to them.   In a semi—or crescent
shaped circle the psychics stand.   The apartment is ex-
actly in the centre of a court which in turn is in the exact
centre of the Temple.   The shrine or Altar that is formed
like a brilliant Sun is in the exact centre of the apartment.   The psychics stand around it in line forming a
crescent.   The master or ancient is in the centre.   His
Oracles and Sybils come next in order—on either side according to their Antiquity.   The apartment is brilliantly
lighted—not by external light, but from the luminosity
that pervades and infills all things, and all beings upon
this Zone, but the Sun shaped Altar is more brilliant
than all else.   The members of the seance stand in
silence, for the vibrations of each one that in the spirit
countries are audible to the quickened ear, make no
sound.   Silence for a period into which each soul enters,
and from which comes a grand uplifting power.   Then
comes a sound as of many voices, in sweetest harmonies
of song, which in turn dies into silence.   Then comes a
radiant light more beautiful than pen or tongue can depict, and standing in the centre of that wonderful light,
a most impressive personage is seen.   It may be the form
and face of a male, or perhaps of a female.   Either are
likely to appear—and sometimes two at once are seen.
These are beings from upper worlds, perhaps the tenth or

twelfth in space.   Possibly it is a planetary being who is seen—a spirit resident of Mars or Jupiter.   Whoever it is, it is an exalted character.   The visitant is welcomed. Presently, beautiful colored lights—like stars are seen to pass from his brain into the atmosphere.   The chief of the seance knows that these lights are words, that they compose the language of the visitor.

He interprets them into a message of truth and wisdom. Presently the lights increase in rapidity producing musical sound in the atmosphere.   One after another of the members of the seance are affected by it—they fall into a trance.   Yet they are conscious of the surroundings. While in this state the upper region is open to them, and visions of the spheres pass before the quickened psychic sight.   The chief of the seance does not become entranced. But while his attendants are in this condition, he makes passes over one and another of them, and as he does so the scene which that particular one is witnessing is pictured in beautiful colors on the air.   Which in turn is indellibly traced or photographed still in colors, by the light of the sunlike Altar, upon one of a series of satin like sheets—though almost transparent in their snowy purity, that are let down from above between the sun altar and the air picture, as described.

Each vision which is being seen by an entranced seer, and which is thus imprinted, upon the air and transferred to the snowy sheet, represents some wonderful portion of the upper spheres that really exists, it is an illustration of real life beyond.   It also symbolizes some grand truth, or conveys some lofty lesson, and it is the business of the chief of the seance, to show these pictures to the pupils in the temple, and to the entire people of his district if they choose to come, and explain each detail to them, that they may not be ignorant, of what they can aspire and unfold to, and of what they need to know of human possibility and development. Each sheet which is of large dimensions, as it receives the imprint of the picture, is again rolled up and it again disappears from view, rising into the air as if by magic, but really borne up to a resting place near the dome of the Temple, there to remain until needed in his public work by the teacher.

As the sheet has appeared at the will of the Ancient, so does it disappear, to be recalled again by the same process of thought, when it is to be revealed to the schools.

All the while that the mystics are entranced, the glorious visitor from beyond is visible. He stands in mid air, the light streaming from his brow, the master of the seance who is controlling affairs, and under whose influence the chief of seance acts. It would be useless for the light, the air, and the will of the Ancient to attempt to photograph this visitor from another world—success would not result, for although this mental photography has gained wonderful powers in the seventh world, yet, it is not there sufficiently perfected to photograph a supernal presence, or a planetary spirit, something of an impression would be transferred to the sheet it is true, but the result would show simply a sunlight disk or light upon the snowy surface—a light in all the brilliancy of natural coloring and beauty but not as a human form.

Such scenes as this described frequently occur. When the mystics have each performed their work they awaken from the trance and one by one glide out of the sanctuary.

The Ancient is alone some of the mystics who were with him, may or may not have been beings of like antiquity with himself, but he was chief in point of wisdom, power and knowledge. He is alone with the celestial presence who is his guide. The Ancient draws near unto the altar, he becomes swallowed by the light of his visitant, is absorbed by the glorious intelligence, and is borne by that power from the Temple to the upper regions, where for a period he enters into direct communion with those who are far above himself. When he returns to his work, in the world that has been assigned to him as his particular field of labor, he is refreshed and strengthened and he goes forth among his pupils, as one who is glorified with an internal light and power such as no tongue can describe. There are other seances also held in the temple, each of which has its particular manifestations of psychical power and wisdom. At many of these one of the female Oracles or Psychics who have been engaged in such occult work, for centuries in one world after another is acted upon by the higher forces and beings of superior states, and through these avenues revelations of planetary life are made. The people of earth who may—because of their own psychic unfoldment and intuition—sense the truth of these statements and be ready to accept them, may be inclined to inquire why we do not describe some of the scenes which these exalted seers behold in their hours of entrancement on high, or why we do not transcribe some of the messages, or instructions that the oracle conveys to her hearers who are to profit by them. We would gladly perform this task if we could, for it would be a labor of Love. But it is impossible—there are no words in the language of earth, no conceptions in the mortal mind of such scenes, and realities as exist in the celestial heavens, and should we attempt to portray them, not one person on the mundane plane could possibly comprehend. But the people of the seventh Zone have been prepared for these revelations, their growth, progress, experience and education, have fitted them to conceive of and understand those lines of life and study, that open to them as they are prepared for them, although there are many things in the higher realms that they cannot comprehend until they reach them, and these are hidden from them even as the doings of Angels in the seventh Zone are unknown to earth. Every human being in the Seventh—as well as in preceding Zones is expected to learn—all are pupils and seekers after knowledge.

There are no infants—as the term is known on earth—in that world, because these Zones are peopled from the spheres preceding them, but there are many who are but as children in point of wisdom and knowledge—even though all are profoundly learned according to the standards of earth—children compared to their wise teachers and guides who are happy in giving them training and instruction as the years roll on.

XI.

The growth and progress of a Soul in the Spirit realms is a Glorious march onward, from age to age and world to world.

What though a spirit entity may elect to remain upon

any one Zone—whether it be the sixth or seventh for one or ten thousand years, Eternity is before him and he knows that he will have no less time in which to travel on from Zone to Zone than he has in the present or ever has had in the remote past. This is the glory of human Life, that it cannot be quenched and though ten thousand times ten thousand ages roll, the soul entity lives in ever nobler power to do its allotted work.

Stars shine in the firmament of heaven. Puny man on earth gazing upon them with scrutinizing eye, wondering of their mystery, and the secrets they contain, yet a little while back, such a little while, as compared with the measureless years of universal life, man on earth considered the stars, only as points of light set to brighten his pathway through the darkness of night, as so many lamps, lighted in the Heavens by the hand of the Infinite, to give him something by which he might see to go his way, believing that earth was the only inhabitable world, of all the universe, and that he was the only being whom God could elect to save. But the stars glowed on. The wondrous worlds of light moved on, never knew nor cared that earthly man, believed them only candle dips to light him on his way.

Yet these are worlds, majestic worlds. Many of them are planets peopled with human beings of intelligence and power. Others are worlds that have not yet developed the condition for sustaining human life upon their bosom, but which will in time arrive at the unfolded state in which they can sustain a manhood that will in itself be as beautiful as the stars. Other worlds that swing in space have had their day as inhabitable planets, and

are now passing through a period of change, throwing off
eminations of sublimated matter, and absorbing from the
atmosphere, and from the human Tutelaries who attend
them electrical and magnetic elements that will vitalize
them anew and cause them to become spiritual planets,
upon which human spirits who are not of the physical universe, but who have entered the spiritual realm itself,
may find a habitation where they can utilize their energies and powers in constructing new forms and objects of
life for the glory of man.

Others of the great galaxy of souls, who are studying
the laws of Nature, and who come in vital contact with
world builders and planetary guides, have given expression
to the theme of world creation. Yet in these pages which
deal with Tutelary Gods and Ancient Spirits it seems as if
the subject should be touched upon.

We have said that while each spirit Zone, is in turn
partly made up from the finer emanations of the world
preceding it, combined with the magnetic elements of
those who dwell upon that preceding Zone, and partly
from the will and electrical forces of planetary spirits,
that each Zone is when completed and started in space as
an inhabitable Spirit world, presided over and governed
by two Tutelary Beings or Gods, one male and one female,
who are forever united as One. Also that the entire number of Spirit Spheres belonging to the Earth planet, and
the planet Earth itself, also of the spirit planet of which
more will be written later on, are in turn presided over by
the superior Tutelary Gods, male and female, who rule
their destinies and determine their career.

Now it is well for mortals to know, that these superior

Gods are not given any name by those who come in contact with them, and who can receive communication from these higher powers—although the latter are never seen by any Intelligence beneath them in points of intelligence and skill, that is, they are never seen as humans, though they may be psychically beheld as great suns or disks of radiant light.

Yet, these superior Tutelary Gods are human beings and as such, display the skill, energy, persistency, intelligence and power of humanity.

The superior Gods are nameless to all below them, hence we cannot record them by any title or cognomen. The Gods who govern each spirit Zone however are known by names not only to their co-workers but to their subordinates and to the world.

We do not call the Tutelary Beings Supreme, who have charge over the twenty-four Zones of the Spiritual system of words, the spirit planet and the planet earth combined, and over the Tutelary spirits who respectively govern each Zone, and the entire train of workers and servants of these Zones, because above and beyond these are other Tutelary Beings who are still more powerful, and above in point of power, design, skill, and intelligence—One Infinite, Omnipotent, Omnipresent, and Chief Power Mind—of all planets and all systems of worlds, whose body is the Universe whose soul force is concentrated as the Eternal Light in the centre of this stupendous frame. Therefore the Unknown and Unknowable of all Tutelary and planetary existences can alone be called the Supreme Being.

Touching upon and magnetically blending with the outermost circle or zone of spiritual life that surrounds the

planet earth, is the outermost circle or spirit Zone of
another planet that moves in space—that beautiful physical orb that is nearest to earth, and concerning which
mortals dream and speculate.   And so with other planets
in the Universe that are inhabited by human entities.

Each has its various spirit Zones or spheres, one reaching out from another, further and further in the fields of
space and the outermost of each blending with the outermost belt of the planet nearest to it in the heavens.   So
also has each physical planet a spiritual counterpart or
spirit planet, and these spirit planets are all peopled with
workers who labor for others in the great field of action.

The spirits who dwell upon the lower planes of spirit
life cannot visit other planets nor even the spirit Zones of
other planets, but human entities from the higher realms
can communicate with planetary Spirits, and can even
visit other planets and their Zones at will.

Tutelary Gods above the third sphere of spirit life, also
the Ancient spirits who dwell above the fifth sphere can
visit and mingle with the inhabitants of other planets—
Spiritual and physical, some of them can control media
upon the planet Mars and Venus and Jupiter, and so on,
and communicate intelligence as to the planet Earth and
its spiritual satelites.   So it is true that certain intelligences from other planets can control specially endowed
and developed media on this little earth and impart information and wisdom—according to the fitness and the
intelligence of their instruments to receive.

In visiting the various planets, the Gods, and their co-
workers or servants the advanced Ancients, can go at any
time by a mere effort of will and desire, but there are

other well informed and progressed spirits—who are of this modern age, who can under certain conditions visit one or another planet.

The conditions are in part as follows: A company of studious men and women of scientific thought and habits desire—in the interest of study and of knowledge, to visit a certain planet. A guide in the guise of one of the Ancients, who has been over the route before is selected to lead the way, and one or two others of like character may be chosen also to attend the party.

But the guides will have to accommodate their travel to the needs and the powers of the company, these persons cannot cleave the planetary air at will. They may do so as far as the environments of earth are concerned, but not in conjunction with the surrounding circles and atmosphere of the planet —Mars for instance which they wish to reach. There are electrical currents and magnetic currents that are unfamiliar to them—these must be met on their journey through space.

Perhaps they must avail themselves of a period when the planet visited is in a certain quarter, or in certain conjunction with the earth, or of an opportunity when the electrical forces and currents are of a certain degree. But at length just the right moment arrives, the company sets out, each explorer well equipped for his flight, and all happy in the thought that they are in reality to visit another world.

There can be no accident to this party, although it may be delayed now and then in its trip, or buffeted about by contending atmospheric forces, or by electrical impulse that impedes progress but, as the journey is pur-

sued solely by the action of the will under the influence
of the guides, the party eventually makes the goal in
good condition, and in happy mood.

We have known a party of such explorers, to reach
their far off destination in a few hours, and we have
known others to spend two years upon the way, but we
have no records of any such party losing any one of its
members, or its self as a whole, coming to any unpleasant
end.

Spirit cannot be killed.   Its elements even those forming its body and drapery are of such an elastic and resist-
ent nature that they cannot be crushed, nor changed by
any volatile or subtle force, for they are themselves of a
like subtle character, Spirit while constantly throwing off
substance, in the form of magnetic aura, is as constantly
absorbing new elements, and there is neither decay nor
accident in the world of spirits.

When the visiting party has pursued its object, and
done its work—it returns to its own home, strengthened
in power and broadened in knowledge.

XII.

While we have in the preceding pages mentioned the
Spirit planet which is a counterpart of the physical Earth,
and which is peopled by millions of intelligent and progressive human beings—of the same grades of culture,
refinement, growth and power as are the various denizens
of the several spirit Zones from the fourth to the seventh
inclusive—all of whom are either humanitarian, philan-

thropic, and working beings, or being trained as such, we have said but little concerning it, and it is now important to speak more specifically of that world.

We have said that each spirit Zone or belt has been created in the line of needs, by, and for human beings. The first or lower belt is made up from the effluvia and emanations of the darkest and crudest human beings who reside upon it, but this darkened sphere was not of such a character until about twelve thousand years ago, because there were no human beings upon the earth who sent out such emanations as compose that plane. The earlier, Ancient human beings were less savage, more refined and spiritual, than are many of the higher classes of civilization at the present time, although many of the people of ancient Egypt, were untutored and humble folk.

The second sphere is created for those who have arisen out of, and been purified from the elements of the first and so on in order as has been before related.

But the Spirit planet which keeps march in space with its prototype Earth, was created soon after the latter physical planet came into existence.

Tutelary Gods and Planetary Spirits from older planets of your solar system, saw the necessity for a world to which the spirit entities of those human beings who were to find an abiding place upon the earth, not only on Ancient Atlantis, but previous to the advent of that mighty Continent, might pass at their mortal decease. These Gods and Planetary beings had been instrumental in helping to fashion earth, and to set it spinning on its course. The Sun, sending off radiations of luminous substance provided the nucleus for the planet Earth, but this had to be charged with electrical force, and reinforced by magnetic elements from conscious human intelligences and it was the work of the Gods and the Planetary beings—as agents of the one Almighty force and Central Light—to perform this task.   Hence under their administration and labor the planet Earth was formed.   After that physical Orb had rolled and plunged along its course for ages, it assumed an equilibrium that held it more quietly and regularly to its orbit, and the plunging ceased.   It then began to generate forces and to create conditions that would in time provide the life sustaining properties and qualities for human beings.   The planet also at about that time began to generate a finer, more etherial aura than it had previously done, the aura of its former period, being of a dark and dense character which settled around it like a cloud and which in reality was the foundation material for the creation of the first sphere or Zone of human Spirits, which long ages afterwards was created or completed by the emanations of the crude creatures who have successively found upon it their first experience as decarnated spirits.

As earth continued to generate, and send forth a more sublimated physical aura and atmosphere, the Tutelary Gods and planetary spirits applied their forces to the refined elements, adding their own luminous, magnetic fluid to the same, until the nucleus of the Spirit Planet was formed.   Breathing upon this, adding power and force to it, infusing it with their own magnetic life, bringing to it solar substance supplied by the Spiritual Sun— for there is a spiritual counterpart of your Solar Orb as well as to each planet of its system—these Celestial workers labored with the new Spirit Planet until it was complete in form and nature, and until it was fitted for the habitation of progressive, intelligent human—Spirit—entities.

While it is lighted by the solar rays of the Spiritual Sun during certain portions of its existence—yet it is never left in the shadow of night nor is any portion of this spirit planet ever held in the embrace of cold or darkness, when the light of the sun is withdrawn from any part of that etherial Planet, that locality may be said to be passing through the period corresponding to your night, but this continues for perhaps a very few hours. During that period those who wish may sink into a magnetic slumber, which corresponds to mortal sleep but which is not needed for more than an hour or two at a time, because it is so vitally refreshing and uplifting, it speedily restores the exhausted energies, or imparts new magnetic elements to the weary frame.

Those who usually require to enter such a slumberous condition are the adults who have parted with much of their own magnetic force in ministering to others, either on Earth or on the lower Planes of spirit life, and also young children.

While in this slumberous state, the individual is guarded and magnetized, by other beings who infuse into them new force and strength.

But it is never dark in any part of the spirit planet, because the planet itself is so thoroughly luminous that it does not depend upon Sun, Moon or Star for light, nor for that matter for warmth, as its atmosphere is equable and balmy, while at the same time of an invigorating character.

Each object in existence there—tree, flour, mountain or sea, Temple, or dwellings and also every man, woman and child, and such few etherealized forms of animal life, that are upon the spirit Planet, also emit a peculiar light that illumes all the surrounding atmosphere—hence there can be no darkness there.

The interior light is of an electrical quality softened by the emanations of refined magnetism generated by the human beings of that world. The flowers and other products of Nature are principally—but not wholly—nourished and lighted by the magnetism, and the illumination imparted to them by their human care takers and gardners, all of whom are students of chemistry and of other scientific laws.

When the light of the sun is withdrawn from any portion of the Spirit Planet, that locality assumes a soft refulgent light which is beautiful and soothing, and yet as brilliant as is the light of a summer day on Earth, an hour or two before the sun has reached its midday height. Sometimes but not often soft and fleecy clouds gather, and a gentle rain or mist descends which veils the light when all things assume a peculiar roseate hazy appearance, the rain injures no one, it only gives a new sensation of refreshment and cheer, and as soon as it ceases to descend, all things bear no trace of dispoiliation or storm. Many of the older or more experienced inhabitants have no need of sleep or of external food. They find rest and recuperation, in a change of thought and labor for the time being. Occasionally these may retire into solitude for an hour, withdrawing themselves from contact with all forces and people, entering into the great silence of the

Soul, and they may or may not, according to their desire at such times, enter into communication with intelligent beings in the higher worlds.

Fruits of various kinds and quality all replete with nutritive properties—grow in abundance in this spirit world, these supply the principle form of food for those who require such sustenance.

Some spirits also make a drink of Nectar from fruit justices and the essence of flowers. The fruits have no seeds and no coarse fibre, their cuticle is thin and translucent, the entire fruit is absorbed by the human system, its minute waste being thrown off in impalpable emanations from the spirit body, these emanations being taken up by the atmosphere again and transferred to the growing fruit or flowers as nutritive quality. Nature there as upon the Earth is economical and busy, a wise and nice utility marks all her works.

We have said that many residents of the Spirit Planet neither eat nor sleep. Yet they are revitalized and nourished, partly by the elements of strength and nutrition which they know how to absorb from the atmosphere and partly from the magnetic force which they receive from guardian spirit entities of the older races of mankind.

The spirit planet was created and pronounced good, Planetary Spirits guarded it, until it was required for the habitation of human beings from the Earth.

Ancients from old Atlantis and elsewhere ascended to it after their physical disease, upon it they made homes, created temples, founded gardens, laid out parks, established schools, and pursued a national sort of existence.

Here they came in contact with planetary spirits and learned of them.

In time these Ancients became busy again with earth, some of them returned voluntarily to it and became re-embodied, others did not. The earth in the course of ages became inhabited by new races, and the various Zones or belts surrounding it were formed as has been previously narrated.

As the need for more belts or Zones increased, each was in turn created and inhabited, some of the beings who dwelt upon the Spirit planet, removed to the fifth, sixth and seventh Sphere. Others remained in their original home. These can pass at will to earth, and also to any portion of the Zones beneath the seventh. Many can travel to other planets, in company with planetary guides.

The spirit planet is constantly gaining new accessions to its inhabitants from Earth.

But, as we have said none but progressive humanitarian, working entities go there. Little children are often borne there after their mortal decease, to be trained as workers or messengers, but only such as the intelligent guides of that world, known to possess the elements and talents needed for such work.

Other children who pass from earth, are taken to one or other Zone, from the second to the fourth, but none are found in the darker sphere.

XIII.

Although there are many Ancients upon the spirit planet, some of whom have homes in various portions of the higher Zones, and others of whom make their abiding place upon that etherial planet, there are yet many spirit entities who have passed from earth in recent years, while new accessions to the human family in that world are constantly being made, to give room and place to the new arrivals who in turn are to become earnest, and zealous workers along humanitarian lines. Intelligences frequently leave that planet to take up a residence upon the fifth or some higher sphere according to their choice and attraction. Such of the Ancients who continue on the Spirit planet serve as seers, teachers and guides to advancing souls of high intelligence who have lived as philanthropic and patriotic workers on earth within the last half dozen centuries. They also labor in conjunction with Media and reformatory souls on earth who are striving to do a good work for their fellow men.

While there are no little children upon the first and darkest plane of spirit life—all such wee people however humble and even degraded their earthly parentage not having generated dense and darkened elements to weigh them down—there are also no small children, upon and of the Zones higher than the fifth, because from that Zone onward, existence depends upon culture, intellectual growth, intelligent action, and power of achievement, and no one can ascend to the higher realms, who has not advanced in the higher studies of life. But there are numerous children upon the spirit planet, little ones who are borne there at the hour of their physical decease, who have within them the elements of useful service so strongly marked that they can be trained to become messengers, guides or special helpers, and guards to human beings on earth—also as helpers and workers, among the dwellers on the lower plane. It is a mistake to suppose, that any one of your friends or relatives who cares for you, oh mortals, and whom you love, is at once—simply because he has passed to another world,—competent to become your guide and guard. Special fitness and special training is needed for such work, even as it is for the soldier, the teacher and the guide in earthly schools, to have special training and fitness for his calling on earth.

True there are many spirit entities who love and desire to bless you—they sympathize with you in your joys and sorrows and they wish to help or advise you—now the influence and counsel of these friends, may or may not be wise and judicious—even though it be well meant—and therefore it may prove to be helpful or the reverse, according to its kind. Ancient spirits understand this thoroughly and they have much to do in the training of our children and attendants of mortals especially if the latter are sensitives or media who rely more or less upon the guidance of their unseen friends. When a child—from the earliest years upward passes from the mortal clay, it may be taken to the spirit planet, and at once placed in a nursery or kindergarten, or in some school fitted for its training, or be taken in charge by some gentle and loving woman, and in her home be surrounded by all that is beautiful and sweet where its tuition at once begins. Or, it may be borne by caretakers and guides to any one of the spheres—according to its needs from the second to th fifth Zone inclusive.

For instance a child of earth may be born of such progenators that it is stamped with inherited moral disease, it needs the care of the magnetic physician and the tender nurse more than it requires mental stimulus. In the second sphere there may be relatives or friends of the child who have suffered and outgrown the lower plane and are earning their way to a home upon a higher Zone, by their desire to do good. Possibly such an one may take the little one into her pretty home, were it receives the care of the physician who attends such cases, and who comes to his—or her—work from a higher realm, or the child may be taken into one of the nursery gardens or sanitariums, that are established there, and receive the care and treatment that it needs, and under which it thrives and grows, and soon becomes fitted to be borne to the third Zone, where kindergartens and schools of training are established for its service and where a beautiful home awaits its coming. That child on passing into the higher realm, will be taken in charge by a relative or friend who has also arisen to its plane, by becoming prepared for it.

Here it may arrive at maturity, or it may ascend to the fourth Zone while yet a youthful being, that will depend upon circumstances and inherent qualities.

Other children of a higher—inherited grade of intelligence and moral sensibility may pass to the third, or fourth Zone, in leaving the mortal. The children on the Spirit planet, correspond in grade of intelligence and moral consciousness to those of the fourth and fifth Zone, and they may travel on through the spheres from such grades as they become unfolded, if they are thus led and guided by matured wisdom souls.

The homes of the second Zone, are simple neat and pretty—resembling cosy cottages of earth, with blooming gardens and fertile fields attached. The sanitariums and nursery gardens and health retreats, where music magnetism and gentle love are the curative agencies employed. The homes and other structures of the third Zone, are more beautiful, and the gardens more expansive and artistic than the second, and so on. Caste and position in the spirit realms are determined by degrees of intelligence, wisdom, and humanitarian labors, nor can any one attain to beauty of form and feature, and to the lovely harmonies and attractive environments of the higher realms in which the spirit planet may properly be included, unless he is fitted to become a part of them. The Condition, position, and grades of the various realms of spirit life all prove that advanced wisdom and intelligence have designed and created them, and while the highest of all power, Omnipotent and Omnipresent has outlined and conceived the work, the details and characteristics of it have been supplied under the guidance and influence of His tutelary Gods, while planetary beings and—so-called —Ancient Spirits, have figured largely in the work as a whole.

The spirit planet being a counterpart—yet an etherialized one—of the physical earth, resembles the latter in many respects, especially in its natural scenery, of mountains, oceans, rivers, lakes, continents, islands and other bodies of land. Also in its fruits and vegetation though there is no coarse animal or vegetable matter or substance upon that spiritual planet, all is sublimated and refined.

The Temples, schools and colleges are fine edifices, some of them stately and grand beyond description and especially magnificent are the galleries of art, and the temples of music.

Many of the beautiful groves there, are selected as places of assembly for the people, and many times have these spots in natures garden re-echoed to the lofty thought, and teaching of wise and grand spirits of the immortal world. Out of door life is far more prevalent there than indoor life, but each family, or each associative body—for there is communal as well as family life there —has its dwelling and its home, each of which is fitted up according to the taste and desire of its inmates.

There are no churches on the spirit planet, and no special observance of Sunday, each day is considered a good and holy day, and indeed time is not set off in the same ratio of periods there as it is on Earth. On the second Zone there are a few modest churches in which some of the inhabitants gather to worship God, for many of these have not as yet been able to give up the idea of a personal being in the form of a gigantic man upon a veritable throne, around whom they may yet gather to sing His praises forevermore. But as they continue to persist in good works, and are prepared to ascend to a higher plane, they lose this idea of God and eternity in the larger light, and clearer understanding of the everlasting truth,

In the first and darkest spirit world, there are many bigots, Catholics and others—who were so self opinionated creedal in their views that they would willingly have burned at the stake or tortured in the inquisition, all who might differ from them in religious idea. It may take years or even centuries for such positive, dogmatic and even cruel minds, to work out of their old conditions, and in the meanwhile they are as eager to proselyte and to help continue the authority of the church on earth as they ever were. Sometimes these beings do so much harm by their influence over mortals, that they are forcibly taken in charge by advanced intelligences and put under restraint until they can be taught, and led by their own awakening desire to learn.

In the second sphere there are no such bigots but there are many who still cherish a reverence for theological teaching and who delight to join with their friends on earth in attending religious service, and in paying special tribute to the Sabbath.

On the third Zone there are many who still observe Sabbath, Sunday, not for any special religious significance, that it should hold over any other day, but because they believe it is well—especially for morals to set apart one day in seven for a season of respite from manual work, and for the cultivation of the spiritual qualities, and in still higher realms many lovely souls have the same thought.

XIV.

There is method and forethought, design, Order and the wisest intelligence in the work and mission of the highly advanced Ancients. They who have lived in conjunction with forces, and elements in the air, and with the spheres for over twenty thousand years have the power of devination of foreordination to a wondrous extent. They can look forward into the centuries and not only calculate what will take place upon the Earth, but they can use an influence to help bring it about.

The Tutelary Gods who direct the planet Earth, and superintend its general development and career, do not attend to details in relation to the affairs of men or of Nations.

Their agents who are appointed or elected, to attend to these matters, are the planetary Spirits, and the Ancients who work with them, and it is these powerful beings, who utilized their forces in helping on the progress of humanity on Earth, and of helping to bring the world forward to a higher development and growth.

The spirit entities who have a portion of this globe in charge—we will, for purposes of illustration, say America, may determine that it is necessary for its best instruction and development as a Nation, to have a leader born within its domain, who will be ahead of his times sufficiently to advance new thoughts, and high ideas to the people. One who will have the courage of his convictions, and who will stand fast by the truth as it appears to him, and flinch not, and who will at all times advocate the higher principles, of justice and brotherly love, as pitted against the lower forces of oppression and greed. One who will, in the face of martyrdom, go cheerfully to his doom, knowing that in the very execution of a sentence passed by tyranny or bigotry will spring up a public sentiment, that will make for grander liberty to man, and nobler justice, in the control of Nations.

Feeling the necessity, of the advent of such a leader, the Ancients plan and prepare for it, exercising influence and correllating forces to in time bring it about. Per. haps their powers will be exerted, upon one generation after another, before the right conditions will be formu- lated, for the birth of an entity who can and will respond to the influence of the spirit, and unfold the qualities and principles necessary for the completeness of the needed personality. But the powers are set to work, and in time, though it be a hundred years—the conditions are wrought and the circumstances produced, under which the human being is brought to earth. This then is the result of preordination—or predelection—on the part of the master spirits of the upper air. The individual whom they have thus planned for and caused to be born on earth, has in a large measure had his life experience planned for him, his career marked out. But he will have a certain limit of free will also, for he can choose his path and also the weapons he may use and his methods of using them, as well as the schools of training, in which he may pursue his tasks.

In the progress of the ages, many such are born on earth. In America George Washington was one of them. So also was John Brown, whose tragic fate stirred mil- lions of Souls into an antipathy to the slavery system. So was Abraham Lincoln, the man for the times, and for the safe guidance of a Nation. So have been many others more or less known to fame. All have been brought forward at a needed moment, and put into action under the influence of unknown guides, who had set their watch for them.

Thus too is the rise, development, and even decline of Nations foreknown to, and more or less influenced by conscious human intelligences, who occupy the wisdom spheres. Nor does it matter to them if the work or its ultimate results be delayed a century, it is all the same, for they know that eventually the work will go on to success.

While these high Intelligences do have a tender regard for individuals, who are gaining an experience on earth, yet they do not allow the whims or caprice of any one to stand in the light of their work. If one, two, or any number of human beings stand in the path of human progress, they are removed or thrown aside. The good of the greatest number is considered, the welfare and enlightenment of the people is of more importance, than the comfort of the one or the few, and therefore the larger good is pursued, even if individuals are crucified in the march.

From a superficial or even a personal point of view this may seem cruel, but from the higher standpoint that measures all things by the standard of human need, and progress, it is shown to be the grandest of all dealing with the affairs of life that make for the universal, rather than for the individual good.

Let us try to grasp the idea of many exalted human beings who are in point of age as old as antiquity, and in point of wisdom as intelligent as Gods, who are exercising their grand powers as teachers, and guides, to

Nations, and guardians over the weighty matters that affect mankind.   Let us try to comprehend, that in all the universe there can never be the loss of a single entity, and that of all who lived as individualized active human beings even in the remote past, still exist somewhere as potent factors in the completion of a universe, and in the perfection of a chain of common brotherhood. Worlds upon worlds exist for such beings and some of them are so far advanced, in strength and will force that if a world were needed, all they would have to do would be to gather the nebulæ of the starry atmosphere, breathe upon it and with their magnetic elements and fluid increase its bulk, density and power of expansion and fashion it into a living world.

Planets breathe just as surely as do men, animals, and plants, and as they are sent out into the Universe to pursue their course, they develope an intelligence that is implanted within them by the tutelary Gods, and the Ancients who are responsible for their birth and growth. Every planet has a physical and psychical nature, just as has a human being, and while the physical part is unfolding its powers and developing its properties, the psychical is doing likewise, and throwing off emanations of etherealized force that are utilized by the spirit entities in new worlds or in completing other works.

Thousands and tens of thousands of years roll on, these mounting up into the hundreds of thousands, human entities from planets that have swung in space, long ages before earth had ever a conception in the womb of planetary creation, have lived and dealt with the forces of Nature, as easily as the skilled workman, here deals with the tools in his hands. Other human entities, who lived on this planet ages ago, come into contact with those other planetary beings and are instructed by them. The ages roll on, men are born on earth, live an allotted time and disappear. They pass on into the great world of spirit. They are individualized human entities still. But it is not for long that they are known, by the names and titles by which they were known on earth. For a few centuries more or less—they may be, but these soon roll by, and as the human egos, are transported to higher spheres and conditions, the affairs that once on earth seemed so big and momentous to them, now dwindle into the past as of little value.

Even the names by which they were known are laid aside, and may be forgotten. Who, that has been an advancing soul for five hundred or a thousand years cares to respond to, or bears the name of John Purse or of James King, or bear the title of Colonel Brown, or of General Smart? Nay, these are things or links belonging simply to but one stage of existence, they were a part of the garments that the man wore, not of the man himself, and as he outgrows not only the garments, but also the age and the fashion of it, and advances far beyond all that, that had aught of association or connection with it, it slips away from him, as naturally as the dust of the road, and the leaves of the trees, slip away from the soaring bird that cleaves the air in its upward flight towards the sky. But earth children cling to their old Names, habits and customs. They seem to think that they will lose their own identity and individual characteristics if they do not. Yet, as the spirit passes on from the third or fourth Zone of progression to the fifth and beyond, it views life in a broader clearer light, and realizes that the Universe is, after all, its glorious home. Thus do the old time habits and names drop away and the new life, like a garment of beauty, upon which is inscribed the name fitting and acquired that belongs to the higher state, fold the progressive spirit in its strength and light.

Ancient spirits are only beginning to be understood and appreciated by a very few of earths—intuitive—people, although there are many psychics who are imposed upon by the forces who dwell upon the lower plane and who psycologize the sensitive, to see them as if they were high potentates, or Ancient masters of great skill, and craft—and for a time these deceivers, succeed in leading astray their dupes—who believe they are guided by high wisdom spirits. But ere long the mask drops away from the pretender, his psychological power is weakened, and he is revealed-in all his trickery to be what he really is, an obsessing and earth bound being, who delights to mislead his fellow men.

XV.

In these pages we have spoken of the birth of man upon the planet Earth, but there was a period in the remote past when man had not found a habitation and a name upon the globe.

Long ages after the planet had been swung out into space, under the guidance and power of its Tutelary Gods and their co-workers the unfolding and the evolutionary processes of life went on within, and upon its bosom, ere it could be fitted for the habitation of human beings. Yet, during all these ages, there were intelligent human beings upon many of the other planets that rolled in space. These planets also had their spirit Zones and counterparts, that were peopled by spirit entities who were intelligent, powerful, and wise beyond the power of mind on earth to conceive. It seems important that we should dwell a little more closely upon these Tutelary beings, and planetary spirits, than we have done, although these writings are not dealing specially with the older planets of your solar system, or with physical worlds still more aged and advanced, beyond the system that embraces earth.

The tutelary Gods of the Universe, are human intelligences, who have advanced from planet to planet, and Sun system to Sun system, through innumerable epochs and eras of planetary life, and who in their progress and power, have become almost infinite but who are yet subordinate and inferior to the Supreme Intelligence of all Life, as all living creatures are. That is, the high functionaries who, in pairs control the evolutions, growth and affairs in general of a planet—such as Earth, or Mars —and all its spirit Zones, and sattelites, are thus graduates of innumerable planets and systems; but the tutelary beings who in pairs control and guide, only one Zone or belt of spirit life—which is also a world—are not thus exalted, for they are the graduates of one planet, and its twenty-four or more spirit Zones. Yet, the latter would seem to be all powerful and grand to even the most learned and great of earth's people, could they come into intelligent contact with them.

The human beings of other planets who lived and progressed ages before the Earth was born, naturally evinced an interest in the formation and growth of this new world.   Its Tutelary Gods and sponsors being deeplp concerned in its growth, and having themselves planned for and helped to prepare conditions for the advent of human beings upon it, felt it desirable to bring to Earth a human power that through material channels could help to magnetize and infuse into it, a vital force that would partake of both the physical and the psychical as the product of human intelligence and will.   At this time the law of reincarnation had not come into existence for there was no need of it.   Nor had the Soul germs—which as points of light, have been mentioned in an earlier part of this work—yet been introduced into matter upon the -planet Earth, for at that period the conditions of matter had not been prepared for such reception.

Looking over their work in connection with the building of a world. The Tutelary Gods pronounced it Good, but they felt the necessity of planting Human magnetism upon its breast.

The design of the superior Tutelaries, had been to prepare the earth to evolve a conscious, sentient, yet intellectual life upon its breast, through one age after another, a type of life that should develope intelligence and memory, as well as intellectual activity.   Nor was this design at any time abandoned.

Yet it seemed desirable to also produce upon the young

planet a type of human life that would understand the laws of evolution and work in conjunction with the Tutelary and planetary forces for the development of a high and noble type of mankind from the lower elements of the globe.

Thus it came about that certain of the planetary spirits consented to take upon themselves the habiliments of matter and to descend from their high state of vibratory action, into the lower realm of earth, in order to bear a part in its productions and in its development of the human family. Thus under lofty Spiritual guidance, companies of intelligent beings male and female did descend to earth —not all in one locality, but they distributed their members to several different parts of the globe.

Each company was attended by a band of high and powerful beings, members of which constantly guarded each entity who had elected to become a denizen of earth. It will not be possible for us to describe how these beings materialized for themselves bodies, formed partly from the finer elements of earth, but principally from the spiritual elements and substance supplied to them by their planetary guides. But this was done so that each came into possession of a body, strong, powerful, yet graceful and beautiful that was adapted to their place and work. Nor can we describe how each formed his and her own associations and established a home in the wilderness, or amid the waste of waters.

But all of this was accomplished. Each male chose for himself a female mate, or rather, each male and female who had been planetary counterparts naturally came together and established their home. Under the influence,

magnetic emanations, electrical forces and will power of
these materialized beings and of their planetary guides
and helpers, the surroundings of their homes became
changed, the waste of waters subsided, and it formed
only lakes or running streams the banks of which—like
that of the Nile—becoming verdant and full of fructile
life. The land around them lost its noxious vapors, and
became luxuriant, with blooming vines and shrubs as well
as with fruited trees the products of which sufficed as
food for the planetary wanderers who had assumed earthly
garments and come to stay.

And so, on the planet Earth that still swung in
space as an undeveloped, half made up globe, there appeared spots of beauty, and of fructibility upon which
human beings could live with ease and comfort. These
garden spots were miles in length, and gave an area of
great utility to the people who became gardeners and
agriculturalists. Homes were founded, the generative
functions of life exercised, the family relations established
schools and temples reared, and communal systems of
human life commenced. If special creation can be said
to be thus brought to earth, then may we truly assert
that special creation was the cause and beginning of
human entities upon the planet.

Under this planetary law and life children were born
and reared who in turn founded families, and taught
their offspring of the higher lore. In the meanwhile the
planet as a whole continued to evolve its various conditions and phases of existence, age by age passing in which
it produced its successive types of development, until at
last it came to the period, when from the animal the

lower types of human life were evolved, for as the animal forms continued to yield up their life elements, and to pass from the stage of expression, these life elements gained sufficiency of vital force to remain in the atmosphere as pigmies and elementals, and to be attracted to the higher forms of life that nature was evoluting and preparing for earth, as the very lowest types of humanity that could be developed from the earthly state. These elementals in turn entered the human atmosphere and form, and again became living creatures that at length passed on to a higher state, to be reborn as finer types of being, because they had evolved a consciousness and spiritual vitality from becoming possessed by and of the Soul Germ that had been awaiting absorption and expansion.

All this took ages to complete, but in the meanwhile, the families of the planetary residents were busy training, laboring with, and directing, their own kind, as well as with the lower forms of conscious mortal life that were coming to the front.   Thus, was it that, primitive man, before the cave dweller and the bushman, evinced an intelligence, refinement and power that linked him with the Gods, while Earth as a planet still labored, to bring about fitting conditions, for the birth place and the birthright of man, who should truly be a child of the planet these earlier people being children of other worlds.   Yet they too had, perhaps millions of years before been evolved from the lower forms of life upon some other globe, even as had their followers on earth.

XVI.

Thus ages rolled on and the planet Earth continued to pass through one period after another of struggle and evolution. The planetary spirits who had voluntarily taken up an abiding place through materialized forms, in one and another locality, and who, after the manner of physical law, had produced offspring and reared families, passed on to the higher life, not to enter on existence upon the spirit planet which was now in process of evolution, but to go to their own spheres, in the universe where planetary spirits dwell, for they were of that class. As the earthly bodies of these spirits, had been materialized and held for service, by the power of electrical force, and of human will and intelligence, so had they become dematerialize, and their elements dissipated into the atmosphere when their period of activity and usefulness had passed. The offspring of these beings had been trained to expect this result and educated into an understanding of the higher laws of psychic existence, hence they were prepared for any manifestation of spiritual power that might occur, and for any occult events which appeared.

The spiritual entities who had for a century, peopled such portions of the globe, had specially been adapted and prepared by magnetic and psychic power for their use, did not cease their interest in this work when they withdrew from it, nor indeed could they do so, because they, as human beings of great love and tenderness had generated the earthly bodies of other immortal beings, who, as soul germs of purity and light, had been attached to them from the great Universe of Causation, and had become absorbed by these earthly bodies, generated by their planetary progenators, had thus become a part of mortality. Therefore, the parental, planetary entities had, a relationship to and a proprietorship in the human beings, who peopled the earth after them. The mortal forms of those early earth dwellers, were made up of both psychical and physical elements and substance, and thus the soul germs that vitalized them had specially favorable, conditions for their development into intellectual egos. The planetary progenators of these mortals, who had passed on to their reward, for faithful service in infusing into the atmosphere of Earth its first breath of human life and power, and its first vitalizing magnetism for higher growth, now took their place as guardian Spirits and guides to the people of Earth, while also becoming high officers of trust in the planetary life.

Thus the centuries rolled, and the people of earth continued to increase and multiply, as mortal and partly as psychical entities—but as the planet continued to struggle, with the conditions of cold and darkness, of evolution and of growth, it presented difficulties to the people in the matter of maintaining a footing upon its surface, for, through the generations, the mortal bodies that were born, became more and more of the physical, and less and less of the psychical although up to the period which we are about to mention, they still retained enough of the planetary and spiritual elements and characteristics of their ancestors to be entitled to be called the children of the Gods.

All along these years the Earth had been bringing forth one after another of its physical forms of active life, Great Giants in animal form had tramped through the forest, or had crawled their slimy way through morass and mire. Floods and storms had spent their fury, in one portion and another of the globe, leaving horrible devastation in their path. At length came a period of greater cold and darkness over all the land.

The gardens of the Gods where the planetary spirits had ruled and set up homes were submerged, and the descendants of that once powerful race, were swept off the face of the earth, but only to take up a happy and intelligent existence upon the spirit planet, that afforded finer and more perfect conditions for a beautiful home.

Again the ages rolled on, the earth righted itself, and continued to produce or evolute one type after another of animal form, each succeeding type developing a higher grade of consciousness and activity until at length, evolving from the series of preceding animal bodies came a form that had less of the brute and more of the human, for although its coat was of hair, and its limbs like unto animal in form, yet it could rear itself, upon its hind legs and assume an erect posture, while its face became more like the lineaments of a human mask. Time passed and the lower types of earthly humanity were born, living in caves and forest jungles, fighting with the beasts for food and raiment. Yet human beings although not sufficiently vitalized by spiritual potency and activity to assume immortal aspect when death claimed the earthly body.

Then came a higher type of man, born as the product of the preceding age—a higher type in which the life principle of that earlier and rudimentary type was absorbed, and one which made possible the advent of the soul germ into a physical body or envelope. But yet the spirit entity had not started on its upward march. A succeeding, in which still a higher type of humanity must appear was necessary, and this also came. The soul germs vitalized the bodies of the preceding race, were again cast forth into the atmosphere. but while they had not gained immortality as spirit entities, they had gained a certain potency and vitality that marked their progress. Now came in the law of re-embodiment, for in the succeeding age, each human child that was born on earth was simply a reincarnated being of the previous age. Later on, as before mentioned, reincarnation served as it still does—as a convenient manner of earth expression and of passing through mortal experience, for such soul germs who may have been deprived of the discipline or stimulus needed to stamp them as progressive immortal entities, and also for such adult and advanced human Egos who for some practical purpose of one kind or another, desire to take up another period of conscious existence as an embodied being upon the earth.

While all along the ages, such elementals as have evolved out of the animal kingdom but not into the human, have the law of re-embodiment in another direction to furnish them opportunity and means to enter the human state—although of course these only do so in the very lowest form, and they have to pass through successive re-incarnations, before they become intelligent and progressive human entities. Nor can this be done until in the earlier stage of humanity, they absorb the soul germs which evolve intelligence and conscious power.

Thus time went on, until the planetary conditions brought a new opportunity to the Tutelary Gods and planetary spirits who had the earth in charge, to again manifest their wondrous power in its behalf. The human beings now existing upon it were of a varied character and degree of intelligence, for climate, atmospheric con- ditions and various other forces and influences had all produced certain effects upon the human race, as it ap- peared in various parts of the globe.

The cave dwellers, and the forest savage had been superseded by higher types of men, who, under the mag- netic influence of the peaceful and advanced dwellers of the spirit planet were evolving a clearer state of con- sciousness and of mentality. Portions of the earth which had for ages been submerged again reappeared, and in such localities, as had once been the gardens of the Gods and the birth place of their offspring, again rose above the rolling waves, new conditions for maintaining a comfort- able existence were found. Races of people of a genial, gentle, and beautiful character slowly appeared. People who loved to study the Heavens and to think out the mystery of the stars. People who were quickened in thought by the influence of the spirits who watched and guarded them, people who intuitively thought of the mys- teries of being, and the secrets of planetary bodies and their movements.

Then came the opportunity for planetary spirits to again descend to Earth—not, this time to materialize, but to quicken the life germs of physical parentage, and to assimilate with their magnetic forces, taking upon themselves birth, growth, and development as mortal children, but so early manifesting their great intelligence and power, as to be recognized by the community as superior beings, and to be elevated by common consent unto high places. Thus did such high beings as Iberna and Cabul and Diurnes take up their abode, and send forth their influences, forces and wisdom in richest measure to the children of earth.

Some of these Gods also found their soul Mates on earth —for the latter also elected to be born on earth and unions of great happiness and love were formed, from which were born children of power, of beauty and of intelligence, and it happened that the children of these Gods were all males, so that when they grew to manhood, they took to themselves wives from the daughters of men, by whom portions of the earth were peopled for they did not all remain in one locality, but in the course of time they scattered to other parts of the globe.

And thus did the earth and its people, again become vitalized from the realms of planetary life, while the earlier beings who had been born from its breast, not only watched its progress from their spirit planet, but they also became helpers in the work of ministering unto its people, while they prepared the way for their coming to that life which awaited them on the higher planet which was to be their spirit abode.

XVII.

During the darkest ages of physical life, and development on the planet Earth, only brute force existed as an expression of animated and conscious life. This brute force was of course the product of the grosser forms of animal life.

To make the earth what it was intended to be—a birth place for humanity, evolved out of its own conditions and types of power and activity, the evolutionary process of unfolding one type—or one species—after another from the preceding stages of development, had to be established and encouraged.

In order to facilitate this work and growth, and to assist the planet in its mission, the earlier races of human entities had descended from the planetary spheres, and materialized upon this earth, that their human magnetism, might become a part of the atmosphere, even of the soil itself, and thus impregnated with the elements and germs of an advanced humanity, the planet would in the far off ages yet to come, be able of itself to produce advancing types of humanity, long after the planteray spirits and their descendants had passed far beyond.

How interesting and ever instructive it is to a floriculturist to watch and superintend, the development of a plant that is putting forth its inherent qualities of growth to an expression of beauty and bloom, and in experimenting with his plants, hybridizing them and producing new varieties of growth and blossom. The enthusiastic gardener finds a dleasure and an education that perhaps he could not gain in any other way.

So with the builders of worlds, and the supervision of planets. In this wonderful work and mission of theirs they find an expansion of Soul power, and of knowledge as well as a satisfaction that is unequalled in any other school of life. For, they are not only expected to watch, and superintend the growth of their planet, *per se*, but also to look after the development of its animate and conscious creatures, and stimulate them by magnetism and influence to higher and still higher grades of mental growth and of physical culture and development.

Therefore if it is gratifying and helpful to the mundane gardener to watch and train his growing plants, and to apply process or conditions to them to stimulate along new lines of plant life and of blooming power, how much more so is it to planet builders, and to the care takers of worlds to watch, train, guide and stimulate to new unfold ments and achievements the physical and spiritual beings who are in their charge, as well as the globe itself which they control. Far back in the remote ages of Earth's history, ancient Egypt flourished as a land of peace and of prosperity to the humble folk who raised their corn and fruits, and who lived in simple accord with Nature who smiled upon them. Fifteen thousand years ago the country of the Nile, was far more beautiful than it is in the nineteenth Century of the present era, and its inhabitants were an intelligent and contented race. Yet, fifteen thousand years is a modern age, compared with the epochs of time and of human existence which preceded it, epochs in which man flourished as a highly endowed creature, who understood his relations to the psychical world, and wno looked forward to the time when he should pass on to the gardens of the Gods.

In this later age the spirit spheres—or Zones had some of them been created and inhabited, and the Ancients were then busy in building or in preparing, still other worlds for their occupancy and use. For thousands of years human entities had labored, in the psychical world for the blessing of mortals and for thousands of years to come they would pursue their onward way in achieving good for man.

In this ancient Egypt many learned men abode, some of whom were descendants of the race that had lived upon, and become engrafted with old Atlantis. For it will be remembered that in the days of that continent, some of the people had departed from it in company with some of the strangers who had come to it from parts unknown. These who had departed, had sailed to other continents of which there were several, and had united with the people there, producing a generation of off-spring, who in turn begat other generations, members of which scattered to various parts of the world, travelling by ships and planting their homes upon foreign soils. Of such as these were the ancestors of the ancient Egyptians who founded schools, and established a system of education and training in the arts and sciences, that has never been equalled in the history of modern civilization. It is true that the peasantry of that remote age, were very far from being highly educated and refined, but even these were kindly gentle folk, who had an inate refinement, that made them courteous to one another, and a natural intelligence that made them very far from stupid people, while the dwellers in the cities and courts were learned and highly intellectual minds.

Somewhere about this dispensation, the spiritual forces of higher worlds, conveyed to the denizens of Earth, through the agency of their Seers and oracles, some knowledge of Tutelary Gods and of planetary beings, and it was learned that each physical world and its spirit counterpart as well as its various spirit Zones or belts were presided over by a superior pair of Tutelary beings. Also that the Spirit planet, and each of the Spirit Zones, had its own Tutelary Deity, all of whom were under the dominion and in the service of the superior Gods who governed Earth and its Spheres.   It was also revealed to these ancients that, over and beyond all worlds and systems of worlds, One Supreme and Omnipotent Intelligence held sway.   It became the custom of the people at large then to pray to—or ask—the spirit guides and the Planetary beings, who they believed to be present in the air though invisble to them, whether it were for atmospheric conditions that would benefit the crops, or for personal and individual riches or position.   After a while, it became the general habit to pray especially, to the God or tutelary being who seemed nearest to them, and as the One of whom they knew best presided over the first sphere of spirit life, their prayers went out to that entity for they believed him to be a human like themselves, and they could not understand the Supreme One whose body was the Universe and whose Soul is Light.   As centuries passed, the people of earth degen

erated, human passions and human greed, and human
love of power and conquest began to gain the ascen-
dancy.   Man warred against man, and Nation against
nation.   Now were the supplications to the Gods turned
into the special channel of petition for strength to slay,
to conquer and to overcome the adversary.

Up from the human family on earth, then went vol-
umes of dense dark magnetism that entered into the
composition of the first sphere and made it more like a
physical world.   The Tutelary of that sphere, in his
sympathy with the interests and the affairs of earth, became inclined to heed the petitions of certain numbers of
men, to the exclusion of others.   Then came in the spirit
of partisanship, and of favor.   It seemed as if that
special Tutelary had imbibed largely of the magnetic
forces of the earth, and was becoming like unto mortals
in sentiment and thought.

His prejudices and opinions were developed like unto
theirs, therefore He took sides—so to speak—and brought
his force and influence to bear upon certain armies and
people for their conquest and triumph.   Thus through
the manifestation of his power and preference there was
born in the minds of the people a belief in his champion-
ship and in his Divinity.

Many of them let go their idea of superior Gods, or of
a greater Supreme O. M., and adopted the belief that
this Tutelary Being who only had the ruling of the first
sphere of spirit life, but who was powerful in intereference
with the affairs of men on Earth, was in reality an
Omnipresent God, and that he commanded the entire
universe.   Through his chosen Media he communicated

with mortals and as he had become influenced by the elements of humanity on earth he displayed many characteristics of the passions or of the life of the times.

There were those on earth who were so well developed as psychics that this Tutelary being could talk with them, and in their presence he performed many wonders.

Nor did he care to make it known that he was only a subordinate of other higher Intelligences and therefore he claimed to be a high Potentate, and ruler over all the earth.

From this condition of affairs there sprang up a race of beings who cultivated a belief in a passionate, vindictive and jealous God, with whom they believed it possible to communicate. From these people, descended in tradition and legend, accounts of various events in which this sanguinary and selfish God largely figured, and in which he was declared to be the one, Supreme "I Am." From these traditions and legends, descended the conception of the Jehovah that in a later age was accepted as the God of the Jews.

But the work of this degenerated Tutelary Spirit, did not continue through the ages as he had believed it would. For he was under the control of higher Gods, who were watching his deeds and thoughts, and in time, when it was deemed that he and his agents had gone far enough, in stirring up the people of earth, to combat with each other. He was dethroned from his position, and put in charge of high planetary spirits to be disciplined, and educated into new lines and thoughts.

XVIII.

But the truth, or the whole truth concerning this very selfish and human Tutelary did not appear to the children of Earth.

And they continued to worship him, as the author of all life and the arbiter of all fate, and to pray to him to send them good crops, and success in warfare or in any other mission.

A successor was immediately appointed, for this dethroned Spirit, and after a while this second Tutelary, began to exercise a more benevolent influence upon earth and it's people.

But the belief in a jealous and partisan God continued to be handed down from one generation to another, although there had now began to be, traditions and legends of two forces in the air, an evil as well as a good Being. This was brought about because early man felt that he must attribute the sorrow, and sickness and discord, and famine that occurred among his kind to some arbitrary malignant power that sought his destruction, and as some of the oracles had given forth tidings of the fall and degradation of an Angel of light, meaning the tutelary before mentioned who had been dethroned—who had been cast out from the pale of the high and good who had trusted him, it had been taken for granted that this was the evil one who hated mankind and whose power was sufficient to do them harm.

And again the darkness of mental and of spiritual bondage descended to earth—holding races of men in its murky folds and the planet was only lighted here and there by a few wise and highly advanced Souls in each age of men but who were not really of the same character as were those who had been evolved out of the lower kingdoms and tribes of earth for they were Ancients who had ages before passed on to high estates, but who knowing the chill and darkness of earth had elected to return and be reincarnated in order to keep the torch of truth and wisdom burning among men.

We are not transcribing for mortals, a consecutive history of the earth, nor of the origin and development of man upon it nor of the rise and fall of nations—such a stupendous work as that, would not be undertaken by decarnated spirits through mortal Media for at least a century hence—but we are giving bits of history and flashes of knowledge upon the origin of Life planetary and human in relation to the earth its spheres and its people, and these desultory descriptions may now deal with portions of the earth and races of people, that existed twenty or more thousand years ago—or again with happenings of a more modern age since the times of Abraham and the prophets.

Now then do we speak of floods and famines and pestilences and warfares, and all manner of evil things that fell upon earth and its people during thousands of years following the era of Moses and the Pentateuch. Conditions and periods of calamity that swept hundreds and thousands of human beings from the face of the earth at one time—time and time again. People who were born in ignorance and slavery, and who were filled with human passions and darkness, going out of mortal existence, by hundreds aye by thousands, began to fill the surrounding atmosphere of the earth with an aura and effluvia that was darkening and replete with atoms and elements of a purely material character.

Thus did the first or lower Zone of spirit life, assume more and more the appearence and nature of an animal plane, and that thus did it gain to it accessions of spirits who in themselves were crude, ignorant and depraved.

In all the years that have passed since that remarkable period to the present day, this lower sphere of spirits has received daily accessions to its people from the depraved, the vicious, and the brutally ignorant classes of earth's children, so too has it been constantly sending off, to the second sphere beings who have outgrown its conditions, and who have been uplifted to a higher state by the help and magnetism, and teaching of the missionary souls who have borne their light into the darkened places, and have sent an influence and power through the low and darksome places with a purifying breath and a cleansing force.

Thus has the planet earth rolled on until now it is approaching an era when more of beauty will infill it's elements, and when a new psychical force will sweep over sea and shore, until it reaches every home and inspires every heart.

The Century is about to close. The twentieth century of the Christian dispensation is about to dawn. There will be many wonderful revelations made to earth by Ancient Spirits, during the next hundred years. The human family on your planet, during the next century will see the grandest of unfoldments in occult demonstration and spiritual significance.

Signs and wonders will be given by the forces who rule heaven and earth. The human mind will be prepared, to receive that which is to follow in the next era of thought and progress.

But not in the twentieth century will the dawning of the new epoch come, for that will not appear until the year two thousand, from the date of the Christian era, has arrived.

For a hundred years to come, time will be utilized by the Spirit powers who govern earth, in bringing about a condition of beauty and of peace, in which man will learn to sink his antagonisms and to cultivate his spiritual nature, allowing prejudice to die away, and candor to take its place.

Early in the twentieth Century a leader will be born, who will at an early age of manhood, rise in the majesty of developed dignity and wisdom to point the way and to teach the people of that which is spiritual. This leader will be full of a spiritual influence, his magnetism will be such, that people will follow in his steps, and be uplifted by his power. He will be a teacher and a guide, pointing the way to a higher civilization and a grander spirituality. This man and messenger will be a reimbodiment of a glorious wisdom Soul, who for thousands of years has watched over the destinies of Nations, and the affairs of man. His work in this last incarnation he will reach, will be to prepare the way for the dawn of the new era of light and of truth that is to come to earth with the year two thousand, and his work will be well done.

Earth hath not yet arrived at its highest state of unfoldment, ages are to pass in which the higher consciousness, and the grander spirituality are to keep on in developing a nobler manhood, and a more royal state of bea-
titude, in which the ethereal emanations of the race will
go out with fuller force and cleansing power, to sweep
through the first Zone of spirit life and purify it of that
which is crude and dense.

Every thousand years completes a Cycle of development and of experience on earth. During a thousand
years, the subtler forces that move humanity may be
changed, that new conditions and affairs may arise in the
history of men.

When the year two thousand dawns, the world will be
prepared to welcome the glorious coming of the truth
whose light will sweep over land and sea with illumina-
ting power.

We have mentioned a leader who is to be born in the
twentieth century, one who will bring victory to those
who learn of him the true principles of human fraternity,
and when his term of earthly labor is complete he will be
fitted to take a place as a Tutelary guide to one of the
higher Zones of the spiritual spheres. For in the sacrifice
of position and power and grandeur in the higher
realms, this ancient who has time and again, come to
earth as a teacher and leader of men, acquires the power
and right to become a Tutelary guide of a brilliant Zone.

Nor will he be the only great man on earth during the
next century, for what he will be to the people in a moral
and reformatory sense, others will be in the line of intel-

lectual training and scientific lore.   Science will so train her votaries that they will be able to perceive the boundaries of physics and of psychics and to trace the elements and conditions of matter into the realm of spirit, so as to demonstrate to every thinking mind the existence of psychical spheres that are real and active with potential force and power.

In the twentieth century one great poet, one great master in music, one artist of immortal fame, one brilliant light in literature, one magnificent singer, one splendid Scientist will rise to fame.   There will be many fine poets and musicians, and many learned minds in Science, Art, and Literature, but these will pale to nothingness, compared to the one brilliant Light, in each department that will arise.

XIX.

Every concious, intelligent entity, earns his or her own place and condition in the immortal world, for while the different worlds in their formation, have been superintended or built by superior beings, yet much of the material which entered into their composition is provided by those who are to inhabit them. The workman on earth who erects a dwelling, does so from the material furnished him by those for whom it is to serve. So the spirit workmen find much of their material in the same way.

No human souls can ascend to the higher spheres, who have not become fitted for them.

We have told of the work of Tutelary Gods—which is to plan and execute in the building and the development of worlds, to supervise the evolutionary processes of life upon planets, and to direct the countless co-workers or subordinates who are their agents in conducting the affairs, of Spirit Zones or of Nations upon their physical world.

A human entity must be far enough advanced in knowledge, wisdom and power to be a guide and guardian of a sphere or Zone, in order to come into direct communication with a superior Tutelary Being, who governs a physical planet and its spirit worlds. The superior Tutelaries themselves, are in correspondence and communication with one another, for although the planet earth is younger, and far less developed than is Jupiter or other planets of the solar system, yet the Tutelary who controls it is as wise and far advanced in knowledge and power as is any of the controlling guides of the other worlds.

But there are still more advanced and powerful Tutelaries than are these who respectively control but one planet, and its spirit Zones. For every Solar System has its controlling power and intelligence. For instance, A Solar System that has seven full formed planets, and innumerable other stellar bodies, each planet and its satelites physical and spiritual being guided by one pair of Tutelary Beings—male and female Counterparts, has itself as a system. One Celestial Tutelary who is above all the rest, and whose wisdom and power are manifested in the intelligent guidance and supervision of the entire system of worlds.

These various Celestial Tutelaries, are in correspondence with each other, but even the high Tutelary who governs one planet and all that belongs to it cannot see these celestial Beings although they can receive communications from them.

Thus are there grades upon grades of intelligent entities. That they are Divine in the higher state there can be no doubt.

But the conception of what is done in even the most advanced mind of earth, falls very far short of the knowledge of Divinity which the advanced souls in the higher realms possess.

As we strive to convey an understanding of these grand truths to Earth's people, we are conscious of our limitations, and we realize how feeble our utterances must be, compared to the magnificent revalations of these occult laws and sciences that are made to advanced studious minds in the Heavenly lands.

In our teachings of the various belts, or Zones of concious active life that in spirit surround and attend Earth, we are frequently induced to use the word Sphere, yet do not consider it the proper term for our service, nor do we like to use it, for each of these Zones, is really a belt of atmospheric strata that is vitalized by living magnetism, and impregnated with electrical fluids; it is substantial and palpable to all the Spirit entities who reach it, and it is in reality a world within itself.

The conception of sphere in relation to human entities and human existence is not as a locality or a world but as a condition.

We know that every ego creates his own sphere, whether it be one of light, purity and honor, or one of darkness, impurity and dishonor. That the sphere created by an ego may be one of Art, Music, Literature or Mechanics. Thus in spirit we define a sphere as a condition, yet it is a convenient word in the definition of worlds, and we employ it in our mention of the several belts or Zones of spirit life as describing a circle or girdle of living substance which in itself constitutes a habitable world.

But whatever phraseology we construct, or whatever language we employ, that belongs only to the people of Earth, we are limited and handicapped, in methods and powers of description and of analogy, for the life and the labors of the Tutelaries, the planetaries, and of the Ancients, are such as cannot be adequately depicted by any system of transcription and illustration which Earth affords. Therefore, only a dim conception of these subjects can be awakened in your minds, although we deem them important enough for this.

Having assured our readers that what we teach them in these writings can only be but a shadow or dim outline of the Glory of the eternal realms, we now proceed to speak of the entire system of planets of all the heavens as one stupendous Whole. As far as we can learn of this, we understand that the entire universe, with its Grand systems of life and power, is the frame work or vehicle of expression—hence the Body—of the One Omnipotent and Omnipresent Intelligence, which is the Infinite and Supreme Life and Light. Of this omnipotent we are told no Intelligence has seen nor learned in fullness, for none but they who were like unto Him could behold or know His Being—none are thus Infinite and Omnipotent, and could even the most powerful of Tutelaries approach Him in grandeur and power—that entity would be swallowed up in, and become a part of the Infinite Light. As the Tutelary Gods from the lowest to the highest are in the Universe, and as none can see the entire Universe at one time, so none can behold the Supreme Omnipotence at any time though those who understand something of IIis Nature and Life, can behold Him in part at any hour.

In the light of the everlasting truth, the common conception of Deity in the mind of man is as crude and feeble as is the glimmering of a candle dip to the glory and refulgence of the noontide sun.

Yet the human mind is expansive, and the wonders of the Heavens are opening out to man, with greater splendor and power as ages roll. Nature works through the laws of correlation and of distribution whether these laws operate in the physcial Universe or in the psychical structure of humanity. It is through the correlating of elements, atoms and forces in the human entity that the wonderous magnetic substance is created and through the force and activity of electrical impulse and will that this magnetic aura, that is substance, is set in motion, and distributed in such manner as enables it to assist in the composition and the construction of worlds, or of other forms of objective life.

Planets are but the materialized forms of creative will and intelligence, they are brought into objective form by the power of Mind, Star-dust and sublimateed matter might float in space through endless time streaming onward in waves and ribbons of impalpable dust or light, but without the potency of world formation, did it not receive the energizing stimulus and the breath of life which the energetic forces of Intelligent design and skill infuse into it.

The elements of soil, water and atmosphere produce upon the earth a certain form of plastic clay—it is the material provided by nature which can be moulded into form and beauty by intelligent skill, but unless the human entity who understands the possibility of that substance takes it in hand and moulds and fashions it according to his will and taste, it will never become a model of statuary, or a useful utensil for the potters service but will remain as formless clay until the winds have dried and hardened it beyond all power of service to the artisan or the artist.

So nature in her Solar and Stellar Laboratories provides and sends forth the material which in the hands of the intelligent artist and designer may be moulded into worlds, but if it is left to its own devices and instinctive motion, it will only stream out in lines of useless dust.

As the potter adds to his clay just the right degree of heat or moisture, and infuses into it some foreign elements as coloring matter, or other substance in order to bring to it just that tint of beauty or condition of durability that he desires.

So the intelligent world builders add to the solar and stellar nuclei whatever of foreign substance they desire in the nature of human elements and magnetism to bring it into the right condition for its formation into a planet, and for its ultimate utility and power. And thus, order, design, skill, intelligence and active power are the necessary adjuncts and agencies in the producing of planets from solar and stellar material, and no world can be created or brought into form, and into the potency of life unless it has been fashioned and breathed upon and received the life elements of human entities.

Grand and glorious revelations are made to the Ancients by the Tutelary Guides who send communications to them concerning the building of planets, and it is from these Ancients who receive the revelations, that these fragments of knowledge now imparted to earth people, are gained, for the Ancients desire to bless the people of earth, with truth and Knowledge of Life.

XX.

We now return to the earth life of the Ancient people who dwelt upon the planet earth thousands of years ago. Those intelligent beings who like Iberna, Cabul, and others with their people who for many ages have dwelt in glorified realms, and who have imparted to the earth and its spirit spheres magnetic life and light from their own exalted souls.

What we have learned of the earthly existance of these people has been gained from their own revelations, in part, and partly from the historic records of their times and events, which are preserved in the Archives of the Spirit world.

For in the higher realms histories of all the races, times, Nations and important events of Earth have been compiled and preserved, and it is only for lack of fitting instruments through which to impart information of these records to earth's people on the one hand, and for want of

preparation on the part of mortals to receive these revelations, on the other, that they have not been given in
more than fragmentary and desultory snatches as yet but
the centuries will yet bring an epoch when full revelations will be made.

At various periods in the earth's history, Continents has
arisen from the depths of the sea, offered their conditions
and facilities for habitation to the tribes and races that
were to populate them, became the seat of Nations and
dynasties, lived their life, and finally disappeared from the
face of the Earth.

These Continents produced, each in turn a race of prehistoric people, and it is with some of these that our writings have to do. Of these, upon a continent that reached
the height of its grandeur and power—as well as its
beauty and bloom, about or nearly trwenty six thousand
years ago, numbered a race of people who were gentle,
simple folk, full of loving service to each other.

A kindly intelligent people, whose minds were studious, and who were disposed to listen to the occult teachings that came to them in the whispers of the night.
When Nature hushed her myriad voices, as if in obedi-
ence to the divine behest, save when the gentle lapping of
waves or the soft stirring of leaves told of the ceaseless
rythm and motion of Life.

Beneath the sleepless stars, in the silent watches of the
night, when balmy hours odorous wit the breath of flowers stole quietly away—these people held their councils,
and their seances in which the wisdom of the higher
realms was evoked, and from which instruction and guidance in all that pertained to human comfort and spiri-

tual elevation was derived, sensitives—or Oracles, Seers, Prophets and Teachers dwelt among these folk. Through and by these Media of communication between earth and eternal worlds many truths were revealed and much of practical instruction given.

Astronomers and Astrologers from other planets communicated with the people revealing wondrous lore concerning the movements of planetary bodies, their period of conjunction with one another, their influence upon each other and upon the affairs of men, and much upon kindred subjects that may not be revealed to Earth for a hundred Centuries to come. Artisans from the higher realms came with their knowledge and imparted to certain minds information of various kinds that enabled those who became skilled in their craft to produce wonderful and beautiful effects for the convenience, utility and comfort of homes and people, effects in pottery, in wood work, in art.

Scientists who had already lived ages in the spiritual realms of the Universe, brought their wisdom down to the people of earth and a knowledge of chemistry and of alchemy was gained, that enabled the Adepts to produce precious stones of every color and brilliancy, beautiful metals of purest quality and fineness, and other rare material for the adornment of persons, homes and Temples such as the world has never since beheld.

The engraver's art then became an exhibition of wondrous skill, and traceries and carvings of great beauty decorated the public halls, Schools and Temples, while one and another branches of handicraft marked the progress of a skilled and refined people whose nicety of perception, was not satisfied with aught but the highest production of mechanics and of art.

In that period labor was considered noble—no thought of degradation of any kind ever entered the human mind, as applied to industry and its natural and inevitable processes and results.

Labor was dignified—for in it's pursuit, the grandest achievements of mind and muscle were brought into objective life.   By its expression human energies were increased, human skill advanced, human usefulness exalted, and human progress assured.   The scientist, the artist, and the artisan stood side by side in the estimation of the race as co-equals in all that make up the dignity of man. No worker was or could be despised, and the man who was set to dig a ditch that the gardens might be properly drained, or to tunnel a path that the mountainous regions might be made accessible, or to build a house for the comfort of a fellow being took as much interest in this work, and imparted his life elements to it, as did the artist who produced beautiful paintings upon the walls of a Temple. The sculptor who carved wonderful statuary from the stone quarried from mountainous retreat, or the alchemist who transformed the baser material of earth and air into the delicate metal that might pay the ransom of a King. Through labor the child or the man found an education and an experience such as he could gain in no other way, and the good of one was considered the good of all. While the welfare of the race was of individual concern and interest.

The schools of that period were of the highest culture and grades of instruction and training—they consisted of various departments of tuition, branches of knowledge in the arts, sciences, mechanics, industries and in the intellectual lines that were pursued by the pupils according to age, adaptability and temperament.

Each intelligent child at the age of four years, was taken from its home into a school presided over by gentle and wise women, and by male instructors of judicious character, benevolent tenderness, and learned minds. The pupil was given the care, training and influence it required for the unfoldment of its best talents and the development of its individuality, each child was expected to cultivate some talents in art—either of music, singing, painting, sculptory, architecture or design of one kind or another, for all of these were included in the arts. Each child also received a thorough drilling in the fundamental branches of liberal education, and was given instruction in one or more branches of industrial labor. Therefore every individual—male or female—became well equipped for the arena of active life to be entered at the adult age, therefore too, every person was a worker in one department or another. There were no drones, nor were there any consumers to help eat up the general store and substance, who were not also producers in the great field of work.

Hence the supply was always equal to the demand, and a system of equity was established and maintained that made the life of the age worth its living, for all who were born upon that continent, every workman of whatever grade of labor he pursued, he was an intelligent man, he not only knew all about his line of work and how to pursue it to the best advantage, but he also knew of the entire merits of the industrial system, he comprehended economic principles and understood the foundation sciences of mechanics.

Nor was his knowledge and skill confined to his special line of effort, for although he might be a hewer of stone or wood or a drawer of water, he might also be a skilled designer, a noted musician or a fine artist, and it was no uncommon practice for the artist or the scientist to leave his tasks in the schools or in the temples and exchange places for a period with the worker in the fields or the quarries—each of whom were enabled and fitted to fill the position thus taken.

Thus upon the ancient continent was labor dignified, and thus was it understood as the heritage and birthright of man, which by its intelligent use might prove the stepping stone to divinest things; for as the mind is trained and strengthened by practice in intellectual and scientific lines of thought and industry, so is the body strengthened and made more powerful by exercise in various branches of manual labor, while also, from this effort the brain is increased in power, and the mind sharpened to grander expansion.

This truth the ancients understood, and in ennobling labor they struck the key note of the grandest and noblest chord of an harmonious manhood that the world has ever known.

XXI.

Upon that ancient continent the inhabitants lived long and useful lives, there were no early deaths—except such

as we shall presently mention—nor were there any sickly
nor debilitated beings. Disease was not inherited,
because if there happened to be any one male or female
who showed signs of physical debility or infirmity which
was seldom the case—that individual was prohibited from
forming conjugal relations, and of exercising the functions
of procreation. Occasionally a child would manifest signs
of mental idiocy, or of physical infirmity, that the wise
Doctors and Seers pronounced incurable. In such a case
the little one was put painlessly into such a sleep as would
enable its spirit to pass out from the body, such a spirit
would be only enwrapped in a sort of fleecy substance, and
be retained in the magnetic aura of one of the seers or
wise ones, until conditions favored its rebirth, in another
mortal form, that could give it better facilities for the
growth of intellect, and the developement of physical
power than the earlier one had done.

The ancients did not believe in allowing an idiot or de-
formed being to live for years in its crippled mortal
form, but they did believe in giving the spirit a new
chance to obtain growth, power and experience through
matter, in a body adapted to its use. Therefore there
were no human monstrosities—of mental or physical
growth—upon that Ancient world. Occasionally however, through some physical accident or experience an ad-
dult would become ill or injured, but the methods of cure
were such that by magnetic and electrical forces intelligently applied, the sufferer usually received health and
tone of mind and body—but if he did not in time, he was
removed to some beautiful retreat where all that skill
could do, and love furnish, would be provided for his

comfort and use, and there he would remain for the bal-
lance of his days carefully nursed, and given every facility
for extracting both pleasure and profit out of life.

In cases too, where it might be thought beneficial for
the incurable to pass into the spirit world, he would be
quietly assisted to do so, without suffering to himself or
grief to his people, for they understood that a happier experience awaited him, but it was very rarely that such
methods had to be resorted to.

There was very little sickness upon the continent, for
the people lived rationally. They understood the value
of a proper diet, of bathing, exercise and pure air, of
breathing naturally and of having their clothing adapted
to the needs of the body.

Their dress was simple and comfortable. Their robes
depended from their shoulders and sandals covered their
feet; circulation was not impeded, and all the muscles of
the body were brought into active use.

Therefore if one should happen to fall ill, he or she
was in good condition to soon recover health. The recuperative powers of the race were good, and the doctors
had practically nothing to do in the way of treating the
sick. Their duty was to teach the people how to live so
as to preserve health.

Even in cases of childbirth there was really no suffering
to speak of, living so naturally made the mother a healthy
women, and she was enabled to bear her offspring in comfort rather than in sorrow and pain.

In giving these facts concerning the early life of primitive man, we do so not only to show what a strong, beautiful and intelligent race inhabited the ancient continent,

but also what a happy, elevated, and intellectual race of decarnated spirits they became after they slipped from the mortal form.

Such a race coming to populate the higher realms, must have generated and sent far abroad an aura or atmosphere that was beautiful in appearance, and composed of elements of strength, and atoms and forces of durability that could not be destroyed.

And from this brilliant and useful material, there must have been formed, not only other spheres of glory and utility but also an envelope of protective power, that encircled whoever or whatever it was attracted to, become a shield of strength and of living might.

Thus were the early ages filled with the glory of an advanced selfhood that even on earth impregnated the atmosphere with a magnetic element and spiritual influence, that even subsequent periods of devastation and of atmospheric disturbences, as well as of human convulsions and catastrophes, of famine, rapine, war, and degradation, through even centuries of time could not wash out or destroy.

Upon the continent with which we deal in this paper—which was only a type of other lands and nations, contemporary with as well as succeeding it—the people lived and reared their altars to the unknown but everlasting God. For while they believed in tutelary and planetary Gods, who could come into more or less touch with them, they had faith in one Infinite and Supreme god, Ruler of all the Universe whom they worshipped as the almighty one.

But even these intelligent and lofty beings, desired a

sign and demanded a symbol of the omnipotent life, and this they beheld in the fructifying sun that gave light and warmth to the earth and in which they found a never failing God—this sun became to them the symbol of beauty and of life, but a symbol only, and never the God himself whom they worshipped and adored.

Nor did they give all their attention and devotion to a Supreme male Spirit—symbolized by the sun.—for while they recognized in the solar orb, the source of light and heat, they also believed in a female being as the consort and other half of the male Deity, and whose potential forces were as important to the earth and its people as were those of the greater Luminary. Thus in the Moon they found the symbol of the female element, and they paid tribute to Luna, the all mother at their altars and shrines, holding a special service of song and music in her praise each month at the full of the moon, when its refulgent rays beamed upon them, turning all their beautiful scenery into silvery light. In the influence of the Moon upon the earth, these ancients beheld a significance, and a sign that could only be interpreted, as a work of beneficience and of good will.

They studied the movements of the tides, in relation to Luna's influence and drew their own conclusions from them. They also learned that the crops were more or less increased or disturbed by the power of the moon, that certain seeds if sown at the full of the moon would germinate and fructify with wonderful activity, and that certain other seed if sown at the same time would decay in the ground, or bear but a feeble crop. They observed also that woman was influenced in her procreative powers and physical functions by the movements of the moon, and they came to the decision that the sun and the moon, bore special relations to each other, and to planetary action as well as to human beings. And therefore they adopted the solar and luna bodies, as their symbols of the dual God head.

The ancients of whom we write, believed all things to have a dual nature, they beheld the male and female principle in all life.

They conceived of nothing without a counterpart or mate, and one of the vital tenets of their religion was that no spirit entity could be complete. Angel, archangel or tutelary—without it comprised the duality of selfhood, man and woman, counterparts or soul mates, two bodies, two soul germs, but one harmonious whole, primitive religion was a beautiful and simple recognition of the intelligent and glorious source of all life and power. A restful faith in the supreme wisdom and justice of that omnipotent one, and a constant striving on the part of man to grow into a higher understanding of divine law, and to personally live in purity of thought and of motive, so as to be fitted to reach the higher abodes of immortality at death, as well as to merit communion with exalted beings, while in the physical form.

Upon the continent of these people there were no wars, for it was not until long ages afterwards, that man came into a state of warfare with his fellows, and sought through physical powers and conquest to raise himself to lofty heights of renown and authority.

To the simple, yet refined and cultured ancients, the taking up of arms against a brother man would have been

purely barbarous and they would as quickly have planned
to murder the infants of their homes, or the wives of
their bosom, as to scheme to slay any of their fellow men,
therefore they had no weapons of onslought nor of de-
fense, for there was no need of such.

Wild beasts disturbed them not, nor in fact were there
any such upon the continent, and when methods of
destruction were required as we have mentiond in the case
of quietly removing the human idiot or imbecile—simple
and painless drugs were used to put them into a sleep,
that knew no waking upon the mortal side.

XXII.

So far we have mentioned a continent as a type of
several that existed—in one ocean and another—from
twenty to thirty thousand years ago, upon each of these
large bodies of land, dwelt in the successive and respective
ages of their history, primitive races, who were in manner
and thought, living, culture, and achivement precisely
what we have described in the few preceeding pages of
this work.

We shall now select the continent of old Atlantis for
special mention, with the understanding that our readers
will consider the people of this ancient continent to be of
just such a class and character as those we have been por-
traying to them.

In the history of the worlds development, there have
been two great and wonderful Continents upon which
advanced human beings lived, which are known by the
name of Atlantis.

But while the smaller and more modern of those continents only existed at a period of from twelve to fifteen thousand years ago, the larger and more Ancient of these bodies of land came to the height of its grandeur and power about twenty six thousand years ago, for nearly a thousand years from that time, the continent as one magnificent Nation, of intellectual and industrial people whose virtue and prosperity have never been surpassed in the annals of human history, flourished, before the catastrophe occurred, which swept the Continent and all its vast and wonderful possessions from the face of the earth.

This ancient continent of Atlantis, was situated in the pacific ocean, midway or there abouts between Asia and America, while the Atlantis of later date, occupied a position in the Atlantic ocean of perhaps a similar latitude.

The old continent of which we write stretched its shining length of mountains and plains over far distances, reaching almost to the vicinity of what is now called the Japan islands, but the entire area of the continent was not inhabited, vast stretches of it consisted of mountanious regions that had not been especially coveted as dwelling places by primitive man.

But the valleys were beautiful spots, fertile and luxurious in beauty and bloom, and chiefly in these valleys man reared his temples and built homes.

Quarries of finely grained and handsome stone of different age and quality were opened and these supplied much of the material of which the more pretentious edifices were composed—while forest trees of magnificent growth, the woods of which were handsomely veined and capable of receiving a high polish, yielded up the lumber from

which many useful and beautiful articles of convenience
were constructed.

Flax, cotton, and a certain quality of silken thread
were produced from vegetable growths, and the looms of
the period were busily employed in turning out fabrics of
delicate though substantial texture and of dainty finish,
rich and beautiful stuffs, colored with dyes of brilliant
or delicate hue, according to the dyers skill and taste,
were woven, draperies of which were furnished for the
schools and temples of learning and of art.

In that age no one ever thought of living in bare and
unadorned dwellings, or of placing the pupils of the
schools in departments or rooms that were not hung with
lovely draperies and adorned with choice works of art.
The beautiful in nature as well as in human achievement,
were called upon to pay tribute to the wants and taste of
man, and the schools and temples were especially advanced with all that could delight the eye, and appeal to the
aesthetic taste of the people. Man in his primitive simplicity believed that life signified growth and development, and that the wealth and beauty of the kingdom
were meant for common daily uses as an educational factor in the unfoldment of the spiritual nature, hence he
believed in bringing all that could be obtained from or
fashioned by skill out of the products of earth, wood,
mountain quarry or field, for the convenience and admiration of humanity, into practical daily sight and service.

It was a recognized quality in the human entity, that
the love of the beautiful was an inate principle, and that
in gratifying that love and admiration the finer instincts
and qualities of the spirit were fed, that would assist in

developing the more spiritual nature, and bring it into higher accord with the Diviner atributes of life.

Thus in the schools the students were surrounded with beautiful hangings, fine pictures, noble statuary, and other productions of art or of nature, calculated to impress the sweeter inflences of being, sweet music too was discoursed at intervals between the studies, and vocal exercises under the auspices of skilled masters of the art of singing, were a part of the training of the schools. Thus, there could be no brutal growth of passion and of carnal influences, in the lives of the youth of that period, for the very atmosphere was tinged, with the refined qualities of all that was sweet and good.

Yet the people were by no means weak or pusillanimous in mentality or in character, nor were their physical powers deficient in strength and quality of endurance. As has been said of the ancients in preceding pages, they were all workers, each human being having been fitted and trained for some class of manual as well as of mental labor and in the exercises of brain and muscle, they developed a power of physical strength, and of thinking and of planing capacity that was grand. Nor were they at all lacking in the elements of courage and of mental endurance, for they were a brave and self-reliant people and not one of them would have been weak and helpless, if fate had cast them adrift upon unknown shores or open seas. In attempting to depict something of the life and habits of these ancient people, we quite despair of giving an adequate idea of them, for this modern age would hardly credit the beauty and the simplicity of a people living in the midst of luxury, art, learning, and yet subsisting upon the fruit of garden and field, and wearing the simplest of home made garments when all the magnificence of a land teeming with the grandest of productions was at their command.

But so it was, for while silken hangings and beautiful works of art adorned the schools, temples and many homes, yet the people lived on simple fruits, nuts, and vegetable oils, and clothed themselves in garments of plain yet graceful fashion.

A studious people content to improve themselves by the instruction of the teachers and high priests, and to study the arts and sciences as well as the various branches of industry, not to enhance their wordly position, or to add to their material wealth and prosperity, but that they might become more learned and wise, and fitted to adorn the world of spirit which they were to enter bye and bye.

Many of the inland cities of ancient Atlantis were built in a circular form yet there were a few like Humenia, that facing the sea, were fashioned like a crescent, the two horns of which seemed to dip into the sea itself.

The circular cities were mostly built of a white and shining stone somewhat resembling Alabaster, and of a very durable nature. This stone was exceedingly hard, yet the finely tempered tools of the engraver and the sculptor were able to cope with it and to work out handsome designs for the ornamentation of building, fountain or court, in the centre of each circular formed city was always a lofty temple dedicated to the arts and sciences. also to learning in other branches of instruction and to religious service.

An immense structure with various departments and

many halls and wings each adapted to some special use. In this temple the high priest and head supervisor of all the schools, performed his labors and wrought his marvelous works for the education and enlightenment of the people, and here his learned assistants and attendants made their revelations or taught and trained their pupils and apprentices according to their work and needs. It is a trite and a true maxim that there is nothing new under the sun, for in every department of human thought and achievment what it has been in the past, time and again. Nature has repeated herself not only in the production of planets, and the construction of continents, but also in the history of nations and of men.

Therefore there is nothing at the present day on the earth however magnificent it may be, but has had its prototype in the past—and the same may be affirmed of the crudest, grossest existing forms today.

Art and science in this modern age may be said to lead the world in their generous creations and revelations. Yet in bygone ages, these branches of human—yet divine —revelation and achivement, far surpassed the grandest of all modern works.

In the light of this fact, man will be pardoned in the belief that the world is retrograding rather than advancing in the higher forms of civilization and of human knowledge and achievement; but it is not so, the human family is forever reaching upward to higher and more glorious things, and the progress-- as a whole is onward. Planets in their evolution, developing, one higher type after another of form and species, and one grander climatic condition after another, pass through their periods of life and darkness, storm and shine, so with humanity in its evolution or unfoldment to higher grades of learning and power, passing through periods of seeming decadence and retrogression, but in spite of all, tending onward in the progressive march, to the highest of all culture, power and knowledge, and to the grandest of all heights of sublimity and love, thus is the world moving on, each age meeting its failures and its triumphs for its ultimate good.

XXIII.

The Atlantians spoke a musical language which was free from idioms, and of the purest simplicity. They also dealt largely in symbols, and their artists produced many pictures, that were symbolical of the life and the times.

The high priests of the temples were like kindly and benign fathers to their flock, but they did not lead the people blindly, for each man, woman and child who listened to the instruction of the high priest was encouraged to cultivate habits of thought and reflection and to practice that of research, so that each could be well informed upon any subject and qualified to intelligently discuss and comprehend it.

The high priest therefore was a teacher—not merely a minister of religous service—and his pride was ever to so train his followers as to have them show the highest unfoldment of intellectual and of spiritual power.

The Goverment of the nation was that of the people by whose voice rulers were elected. The chief of the nation was one of great learning and of practical thought and wisdom who was selected by acclamation.   This chief appointed his own advisers who were also individually male and female of high standing, cultured, wise, and refined.   These advisers constituted with their chief a council of twelve, and their appointments always received the ratification of the people, for if any member of the council should be objected to by the popular vote, he or she would be excused from taking a seat in the council and the positon would be filled by the choice of the populace.   The continent was divided off into districts—similar to the states of America—each of the districts was presided over by the high priest of its temple, who had in charge the various offices and functionaries who made up his council, and whom he consulted on all important affairs of state.

This high priest and his staff were all selected and elected by the people, the high priest himself, holding his office through his life, his functionaries holding their positons for a period of five or ten years respectively, with the possibility of re-election for another term.

Every intelligent man in the district was, expected to be capable of filling any high office to which he might be called and it was a point of the education of the schools to implant a knowledge of government and of economic principles in the mind of each pupil.

The system of government was of a parental and a patriarchal character.   The government managed the industries, and through its offices, apportioned off to each of its children his share of work.   All the products of field and orchard were turned into one common store-

house or granery and distributed or disposed of according
to the wisdom and decision of the council, while each in-
dustrious individual received his proportionate share.

Electrical force was largely employed in the various
mechanics and industries of the age and was well understood. Its machinery was delicately constructed and not
of a cumbrous character.

Especially in modes of transit was this electricity used,
and the carriages and ships and air cars were propelled
mostly by this serviceable agent.

The various industries being conducted upon a wise
and just plan of distribution, and proportionate equality,
there were no long days of labor required of any one, six
hours of active work, six hours for study, six for recrea-
tion, social intercourse and holding of seances, for communication with the unseen, and for any pleasant and
helpful entertainment, and six hours for sleep.

This was about the proportion of time held out to those
who were at the working age.

The ancients being a healthy people, and being true to
nature were not easily exhausted of nerve force and
tissue, hence six hours of slumber out of twenty four sufficed to give them all the recuperative force they required.
They had then many hours in which to develope their
tastes and cultivate the personal qualities and friendships,
as well as to study deep subjects, and perform such labor
as was required of them.

And life upon the continent was a happy existence,
pleasantry and sociability abounded, hospitality marked
the family circle, the whole country was filled with a contented, kindly disposed, courteous and intelligent race

that sent out an atmosphere of hearty good will and cheer, and an aura of light and perfume that permeated all things, with its radiance and warmth.

There was no poverty upon Atlantis, no one could become poor and wretched, for there was no injustice, and no discord in all the realm, therefore beggary was unknown. The traveler through any populated district could set out on a journey with no food in his pouch, and with no coin or its value by which to purchase lodging, for he was as sure of being kindly entertained as the kith and kin of the people along his route, as he was sure that he would kindly entertain any one who might be a traveler in his own vicinity. For indeed the people of each district, were as the brothers and sisters of one great family of whom the high priest was the head or sire and his consort the mother of them all. And should any one from other districts arrive they were taken in and cared for just as if they were expected and well loved kin.

The high priest and his functionaries, were all married people. Celibacy was not in order nor in favor on the ancient continent; conjugal relations were sought and upheld. The felicities of wedlock were extolled, and the duallity of Life as symbolized in the marital relation of man and woman was revealed as the sign and token of the great Dual force of being, represented by the solar and lunar Orbs.

In those days there was no inequality of the sexes—man and woman occupied equal planes. In after centuries there arose nations in other quarters of the planet in which woman held superior office and ruled the affairs of state, but in the day of these ancient Atlantians, man and

woman occupied the same station, and each was the com-
plement of the other in wisdom, order, knowledge and
power.

Each high priest had his consort, who was also a person
of dignity, of honor, and of intelligence—so too, every
other officer of general affairs had his female, or her male—
counterpart.

Women frequently filled offices of importance, and dis-
charged their duties with wisdom and ability. Occupying
equal ground with man and provided with the oportuni-
ties for the cultivation and growth of her intellectual and
spiritual qualities, it goes without saying that the children
of such a woman came into being well equipped for the
duties and the experiences of life.

Thus upon ancient Atlantis the beauty and bloom of
paradise reigned. It was the garden spot of the world.
Its people were magnificent specimens of man and
womanhood. It was a golden age in which the higher
proclivities and principles of humanity prevailed. Yet we
are told that even a higher and grander cult and manhood
will reign on earth, and even a more delightful golden age
will come before the planet reaches its height of progress
and begins its decline.

The Atlantians possessed a knowledge of printing or of
transcribing thought into language in such form as to
preserve it for future generations. But the ponderous
machinery of the press room was unknown. Their manner
of printing was different from the modern process. Their
presses were constructed of delicate frame work. Men of
genius and skill were the printers, whose good judgement
selected the news and the matter to be published.

There was little of local affairs printed, but news from the upper spirit regions, through Media was abundant and bulletins of information accompanied with appropriate illustrations of the subject represented were frequently sent out through the district and over the land.

The people living in the cities mostly pursued the mechanical arts and industries or flourished as musicians, singers, poets, painters and scientists.

Those of the outlying country were agriculturalists and workers in wood and stone. Once in seven days the people of all the district came together in and about the temple and the city was filled with the cheerful throng. On that occasion public exercises were held, consisting of preaching, singing, music, oratory and art exhibitions. The service of the day was especially religious, yet it had both a spiritual and intellectual significance that was precious to its observers. On this day of general meeting a feast of fruits, of corn and of unfermented juices, was held at which libations were poured upon a flower crowned altar to the gods and thangsgiving offered up in tuneful voice to the supreme one for the blessing of life.

Corn and rice were raised in abundance upon the continent which afforded much nutriment to the hungry. These were well cooked and eaten with wild honey or with sugary juices distilled from fruits of rich flavor, and were highly esteemed for their nutritive and palatable qualities. Once in six months a special festival was held by the central goverment in the most important—or capital city of the nation, to which many of the people of the various districts journied for it was considered a festival of the gods.

XXIV.

The ancient Atlantians were a noble people, in personal appearance as well as in spiritual culture and in mental ability.

They had no doubt of the soul's immortality for they were in constant communion with the denizens of higher realms, from which much of their own wondrous learning and training in the arts and sciences was gained.

Each of the grand temples of that vast continent, was built around a great square or court, which was paved with precious stone. In the centre of this spacious court, stood a magnificent altar built upon geometrical lines, and of costly workmanship. The altar was composed of rare and brilliant metal which was so highly polished that the rays of the sun reflected from its burnished surface as from a crystal mirror. Upon this altar fruits and flowers were daily heaped in rich profusion, not as an offering to an unknown God, but as sweet offering to the spiritual guides, who directed and attended the people. On the evening of each day these sweet herbs and fruits were burned, and from them there arose an inscense that in its aromatic daintiness seemed a fitting offering to the heavenly world.

The climate of the continent was an equable one—semi-tropical fruits and flowers grew abundantly. A luxuriance of growth—not in rank profusion, but in beautiful symmetry—marked the valleys and the foot hills of each mountain range, the seasons varied but little in temperature and atmospheric conditions.

That which corresponded to a summer of a later period was marked with certain warm winds, that were unknown in the winter season, but these were tempered by the freshness of the sea, while the cooler seasons were devoid of frosts and of wintry snows. So that at all times the flowers and fruits grew abundantly.

There were no heavy storms; each period of time had its sunny hours, and its hours when gentle rains softly fell refreshing earth and sward, with their beneficient dews. The nights as a rule were comfortably cool, and of a clear and beautiful radiance that lent a bewitching glamour to valley and hill.

Crops were easily raised, and the husbandman had the pleasure of receiving bountiful harvests for his pains and labors in orchard and field.

Atlantis was a beautiful world, a continent of power and of loveliness, capable of producing fine products of nature and of humanity. A kingdom that the greatest potentate of all the earth might envy and aspire to rule.

But Atlantis is of the past; thousands of years have swept over her grave in the mighty deep. Epochs have come and gone. Nations have risen and decayed. Worlds have sprung into beauty and spun along their shining track; Races have come upon the scene of earthly action and have been swallowed up in the mighty tide of evolution and of time, but still the waves chant their requiem over the graves of old Atlantis and still her magnificent cities lie buried deep, deep beneath the shining waters of the restless sea. The capitol city of that ancient conti-

nent was situated in the centre of the country, so as to be
accessible from all regions where man abode.   It was a
splendid city, circular in appearance and built up of shining stones, some of alabaster whiteness and others of rain-
bow tinted hues, as is mother of pearl, the public buildings
of this city were very grand and massive and there were
many of them so that the entire City shone in the sunlight—or even beneath the rays of the billowy moon—as if
it were set with magnificent and immense jewels that
shone with an interior luminosity in some instances and
in others with irredescent light.   In this central city were
the chief buildings of government and of national affairs.

Here too were the spacious and handsome residences of
high officials, of priests, teachers, seers and philosophers.

From this part of the world went out many an advanced
character to found schools, or to fill positions of trust in
other localities.   Once in six months—as time is now understood—the high priests and the teachers of each
district journeyed to the capitol and came into conference
with the chief executive of the Nation.   The deliberations of these councils lasted for three days, after which
the visitors returned to their homes and their own charge
to pursue their work with fidelity and zeal. In the capitol
city were many temples of beauty and of learning—but
the chief of these was the seat of government which occupied an immense square, and which was presided over by
the head of the Nation and his noble spouse.

For convenience sake we will call their principle city of
the nation Etruria, and the chief executive may be known
in these pages as Constantia, for we are now approaching
a time in the History of Atlantis when the presidency of

one out of many great and good souls who were identified with the existence of the continent, stands out in the spiritual annals of the times with brilliancy and power.

The period of which we now write was not that of the closing century of Atlantis, but one that preceded it for three hundred years —let us here remark that as the Atlantians as a race were a long lived people, it was no uncommon sight to see individuals actively engaged in their studies and pursuits who had all the appearance of persons in their full vigor of health and strength who had rounded out more than an hundred years.

At this time—when Constantia presided over the affairs of the nation, the chief executive was usually elected for a term corresponding to that of twenty-five years of this modern age but at the close of this term, it was made a law that the president like the high priests or governors of the various districts should hold his office for life. The possibility that any officer could or would do anything to deserve impeachment or deposement never occurred to this simple and harmonious people. The continent had flourished for over a thousand years and there had never been any sign of treason or of mismanagement on the part of any of its officers; and so well preserved was their wisdom and manhood a patriarch of one hundred years was fully competent to guide the affairs of state. To be sure there was but little technical or complicated machinery to be run. Simple rules of living, beneficient laws and spiritual teachings completed the code of Government as well as of morals.

Religion and politics were all one in the sense that a knowledge of spiritual law and its relations to man, un-

derlay the man made law that protected or guided
the nation and its wards.   What was of importance
to the race whether religious or political had an equal
significance to the people, and as harmony prevailed the
country looked like one beautiful home, in which each of
the family members bore his or her part with the utmost
willingness and grace.

During the administration of Constantia, the practice
arose of holding seances at stated intervals during which
high messengers from supernal realms announced themselves, bearing heralds of significance from above.   This
was brought about through the agency of Stancia the wife
of the high chief, whose occult powers of penetration were
of a grand character.

Seances for holding communion with the unseen forces
and intelligences had long been in vogue with the high
priests and various teachers of the different districts, but
the common people had not taken part in them.   But at
this time, Stancia declared that the Tutelary Gods of the
planet had made known to her through his planetary
co-workers, that the time had come for the masses to be
more fully instructed in the lore and the wisdom of the
heavens and that seances were to be held in each community which all who chose might attend, and that she—
Stancia, was to go forth and visit each district, holding or
establishing a public seance in each, and set to work the
forces and intelligent powers of the air in developing
media for the communion of souls.   It was also announced
that planetary beings from six planets were to visit earth
and through her mediumship found an order of great
power and significance which should include in its ritual

the rights observed by similar bodies on other and older planets, and that this order would be composed of all persons who had attained a certain age and degree of experience, who would be willing to specially train and prepare themselves for their initiation into it, and that the order would be the means of developing a class of psychics or seers that would be all glorious in power.

It was further revealed to Stancia by the Tutelary— through his agents, that in the course of a few hundred years Atlantis would lie engulfed, and that all of its people, but a few who would previously sail away to other parts of the Globe, would perish, but that the principles of this ancient planetary order would be preserved, and handed down to future—or succeeding ages and tribes of men, and again and again be incorporated into forms and manifestations for the uplifting and instruction of humanity.

XXV.

According to the instructions thus laid upon Stancia by the planetary guides, the work of opening seances in various districts and in establishing an occult order of such individuals, who had proved themselves willing and competent to observe the special rules of training and culture laid down for them, was inaugurated.

Beginning with the founding of this occult school and seance in Etruria at the grand temple, each distict of the country was in turn visited by Stancia and her spouse Constantia, at which the work was continued.

Then came many of the people to learn of the things which were to be revealed to them through the instrumentality of the first lady of the land, and the order which she established flourished and grew, until it included, in its membership a representative of each family in the country.

In each district the branch of the Order, held it's councils semi-monthly in the temple of the city, and at these regular meetings various exercises were held; among these exercises were sweet musicians and also the sounding of musical instruments that were unseen by mortal eye, but plainly beheld in the hands of celestial visitants, by the psychic sight of the sensitives present.

There was also a communication read by the chief of the Council from Stancia, of matters upon planetary studies, also on spiritual topics that were to be studied and discussed by the members of the order, as well as messages received and transcribed by the Media in the temple directly from the spiritual world, at these meetings much knowledge of life and the Universe was imparted from the celestial realms, and under the auspices of this grand order of psychic study and research. The country flourished as it had never done before.

Once in every thirteen months—which in those times constituted a year, there was a grand meeting of all the branches of the mystic order at Etruria and a general exchange of wisdom and a comparison of labors were expressed.

At this grand mass assembly, Stancia was always influenced to make an address of great length and of matchless eloquence and power that held her auditors spell bound for hours, and no one dreamed of being uneasy or weary at the length of her address.   During the progress of this order—and of the administration of public affairs by Constantia and Stancia, one planetary intelligence from each planet of the six frequently took posession of the organism of their oracle and gave wonderful information concerning the code of ethics that prevailed upon their respective planets.

From these statements it was learned that every one of the planets of the solar system, was even in that remote age inhabited by advanced human intelligences and that it had been so for ages of time. Also that millions of other planets existed in the universe which were inhabitable worlds, and that countless solar systems revolved in space, each one comprising a universe of potency and power within itself, and all filling the Heavens with glory and with ineffable Light.   It was then revealed that the Universe is a circle of light without end or beginning as far as human penetration can learn, and that it is in itself the stupendous framework or body of the Everlasting God, that in its centre glows and burns the soul force of deity from which primarily emanates every soul germ of planetary life.

That each star and planet and sun of every solar system is but an atom of the great body of Infinite power and that the whole starry universe composes the manifestation of the supreme intelligence. How vastly more grand and glorious is this wonderful revelation and conception of deity, than was that of later ages, in which man believed the little earth to be the only inhabited body of the skies, and the stars to be only lamps to

light his way.   That belief that thinks of the abode of
God as in the narrow centre of the theological heavens,
consisting of a white throne upon which he forever sits
in majesty surrounded by angelic beings forever shout-
ing Hosannahs to the lamb.

Oh that we could transport our readers back to the
country of the Atlantians, that they might listen to the
revelations of Stancia, and revel in the beauty of Atlantian delights.   But no, that garden of the Gods is
washed over by the waves of a restless sea, and its people
have found higher joys in the creation of spheres of
Light and of wisdom that are affording homes of lovli-
ness to countless hosts of progressive souls.

In every age of the earth's history, however, since the
time of Atlantis there have been born some human beings
with conceptions of the true nature of the Universe and
of God, beings whose intuitions have led them in the
true paths of contemplation and of reflection in spite of
the crudities of external life, and of the narrow formulas
of faith and creed which error and misconception have
woven for the world.

Chief among the planetary tutors and visitors who
possessed Stancia for a time came a high personage from
Jupiter.—This was a magnificent specimen of manhood,
about fifteen feet in height of massive proportions and
splendid presence, a being whose countenance shone like
the sun and one whose very nature must have seemed
that of a God to the oracles and seers who gazed upon
him.   The planetary visitor did not of course enter into
the aura of Stancia, but he enwrapped and overshadowed
her by his glorious atmosphere and influence and while

she was thus possessed her own countenance and figure became illuminated and transfigured, beyond the power of human mind to describe.

It was from this visitor from that wondrous and majestic planet now called Jupiter, that revelations of his planet were gained—knowledge of its marvelous breadth and height; of its stupendous possibilities, of its people and their modes of living, all of which were very far in advance of even the beautiful life of Atlantis on the planet Earth.

From this Jupiterian visitor was also gained a knowledge of the moral code of this world, but then be it remarked that upon advanced planets, there is in reality no need for any expressed code of morals in the guidance of the people for they are a law unto themselves, a law of purity, goodness, and of wisdom.

The language of the Jupiter folk is a universal one but very different from that of any tribe or race that earth has ever produced. But the planetary instructor from that world knew how to adapt himself to the understanding of the Atlantians and could we but reproduce for the people of earth in these modern days the sweet significance and lofty utterences of speech and thought of Atlantis, it would be a pleasure to express the moral ethics given from Jupiter to the people of earth. Yet if we do our best we may succeed in making an impression upon some of the intuitive souls who are ready for the truth, as translated from the Atlantian transcription of fragments of Jupiterian revelation, we may quote as follows:

"Of time there was no beginning, neither can there be

any end. The universe of space always was, and ever shall be.

The essence of planetary life dwell in the bosom of the infinite power, and ever springing into form at the breath and will of God. Eternal causation rules the Universe. Its potency is forever bringing into life new forms of Being and of activity. God ever has been and ever will be.

The finite mind comprehends not God. The genesis of worlds and of men may be revealed, but there is no genesis of the creative force of God for that which ever has been and ever shall be can have no generation and no history.

Birth, Growth and decline can have no part in the life of God. Man is but a globule in the life current of the infinite, globules yield up their force, their forms decay, but the potential qualities live forever in new forms and revitalized activities.

Man lives forever in soul power and force. He is a microcosm of the Universe. Within his frame are all the wondrous arteries and tissues and fibers and powers of the great whole, in miniature. What the living breathing form is to the soul the universe is to God, without the soul essence and light, mans organic frame would have no power, it would not hold together as an organized form.

Without God the Universe would be void of force and light, it could not hold together as the framework and setting of worlds.

God is light, God is force, God is intelligence, God is life. Souls march on from world to world. No human ego is destined to spend an eternity on any one planet and its spiritual zones.

Ages may roll while the entity is gaining experience and unfoldment on one or another planet.   Thousands of years are but as so many whisperings of the sea of the spirit.

Ten thousand times ten thousand years may pass while one being is gaining his lessons in a world, yet shall he move on to higher planets, move on to the music of the marching spheres, forever gaining new distinctions, forever reaching unto new pathways for knowledge and power.

Eternity swallows up a million of years and the soul moves on into new conditions, it sweeps on reaching the planets that burn in celestial glory along the highway of the stars.

On, on, from zone to zone, from milky way to milky way, forever on, not alone, but attended by companiable hosts whose march like his is forever on."

XXVI.

Such are the fragments of what the Atlantians received in the days of Stancia the oracle.

Here too is the synopsis of the unwritten moral code of the Jupiterian, a code from which there is no appeal and no deviation a code unwritten but accepted and practiced by all people.

"God is light—therefore shall we seek to walk in the

light". "God is Light—therefore shall we abide in the light, and shall the light abide in us."

"God is Light, God is in the soul of man, hence light is the soul. Therefore shall we unfold the soul by wise living and noble practice, that the light may shine from and through our bodies to illuminate all our path."

"God is love—therefore shall we seek to walk in love."

God is love therefore shall we seek to abide in love, and shall love abide in us."

"God is love, therefore shall we behold the face of God in working through love for the blessing of men. Therefore in cultivating love in the soul and sending it forth through light, shall we feel the presence and the breath of God."

"God is wisdom—let us seek wisdom and understanding then shall we seek God and his glory. God is wisdom therefore as we grow in wisdom shall we be growing in the likeness of God which is beauty and power!

God is life—in life we live, and breathe and move, in life unfolded by the highest principles of light and love and wisdom. In life we are centerd to the vast possibilities and potencies of Causation and Expression. In life we are one with God."

"Light dispels darkness, love creates harmony, wisdom begets power, life contains all potentialities. Life is God. Man is in God, of God. God is eternal good. Man must work for good, grow for good, live for good. He can only generate light, love, wisdom and live the true life who is good."

Such an imperfect transcription of the moral code and unwritten creed of the Jupiterians as given to the people of Atlantis through Stancia, the beautiful wife of Constantia the good and wise chief magistrate of a nation of peaceable and intelligent human beings who flourished in the Eden land of Earth twenty five thousand years ago.

It was the custom of the Atlantians to chant invocations to the high celestials who came to them from supernal spheres and planets, inviting them continued ministration and expressing gratitude for their attendance, and it also became the practice of the planetary visitants to respond in chanting measures and rythmic tune to the salutations of their children on earth.

Occasionally there would be a public procession through the beautiful city streets, a procession of just numberless vehicles—flower wreathed, filled with white robed maidens, each wearing upon her breast a miniature symbol of the sun—then followed on foot a column of young lads clothed in sky blue robes, wearing upon their breasts a symbol of the ancient, order a crescent and star, after these appeared the president and his spouse in yellow robes gorgeous and silken, he wearing the emblem of Sol —and she that of Luna. This pair and their council rode together in a snow white chariot wreathed in vines of flowers—not wholly like the pretty "morning-glory" of to-day. Their chariot was horseless—as were all the vehicles—and was propelled by electric power. Following these dignitaries came men of youthful vigor yet bearing the marks of experience who were handsomely gowned in the robes of dainty green that symbolized the fructibility of the earth and its luxurious growth. There were musicians who discoursed sweet music along the route, they bore wind instruments of golden material upon which they blew with precision and skill. These men also wore insignia which symbolized the bounties of nature, and which were of beautiful design.

After these marched the high priest in robes of office—snowy, girdled at the belt with golden cords—these priests were decorated with a golden miniature of the globe. Following them were squads of teachers, philosophers, artists, musicians, scientists and seers, some on foot and others in vehicles and all wearing their appropriate robes and insignia.

Nor were the riders and footmen obliged to walk or ride according to caste, but each chose his own method of locomotion and many of the highest officers preferred to walk while some of the humblest citizens rode in vehicles in the great parade.

Among the bands of musicians was one that presented a spectacle of great beauty and which pleased the spectators very much, this was a band of harpists and players upon small stringed instruments—the harps being small in construction and easily carried, this band consisted of a number of lovely young women representing the sea, each of which was clad in robes of blue crested with foamy white, or of sparkling green tinted with a purplish bloom and flecked with purest white to resemble the waves in their varying lovely tints.

The procession thus mentioned was one in honor of nature, the Goddess of life, and was usually formed in recognition of some unusually bountiful harvest or of some event in nature that bore greater significance to the world than the customary events or seasons gave.

But to return to Constantia and his beautiful wife—

Stancia the oracle through the advices which he—the President received from the upper world. Constantia inaugurated a system of public exercises that served to bring the president and his council in direct touch with all the citizens and which only served to cement the bond of friendship and affection existing between them all. So equable was the climate that the people were enabled to live out of doors, and Constantia caused an immense area of land to be converted into an open park, laid out in squares and beds of flowers, groves of trees and banks of verdure, and adorned with fountains and statues of grace and beauty. To this park it was the custom of the chief executive to repair toward the close of day and to hold intercourse with any of the citizens who might wish to converse with him upon any matter whatsoever—for there was nothing that could be of the slightest interest to even the humblest—if there were such—of his people that was considered insignificant to the great and wise man.

It was upon one of his visits to the park that Constantia, walking quietly by himself in a grove of lofty trees the foliage of which made shadows all around him, that he beheld a man of gigantic presence rapidly approaching him, as apparently out of the very air.

This visitor had a shining countenance and appeared as one who had come from some celestial world, for though the higher class of Atlantians such as the high priests and oracles had really been born of superior parentage and were themselves but reincarnated planetary beings and hence were of shining aspect and destinguished presence, yet there was something about this visitor that set him apart from the people of the continent and which surely designated him as a being of another and a more advanced race.

As the gigantic man approached Constantia observed that he was not walking, but was rather gliding in the air, and that he carried in his hand a scroll of shining white parchment that glittered like a roll of light in the fast increasing shadows of the twilight. For it was now the close of day—most of the people had gone to their homes, and many of them were busy with their dainty evening meal of cakes and nuts and fruits, Constantia was alone, for at this hour Stancia was always closetted in the temple with her celestial guides and usually in a trance condition in which she travelled about in spirit through many spheres visiting other worlds and meeting with beings of planetary power and light.

Thus Constantia was alone to meet his celestial guest; the latter continued to draw nearer and in a moment descended to the earth and stood before the ruler—or president of Atlantis. But in the presence of this august being the great president seemed no larger than a child or at least a dwarf, and the visitor was obliged to seat himself upon the ground in order to bring his face on a level with that of the Atlantian man.

XXVII.

The entire face and figure of the celestial visitor shown with a strange and brilliant luminosity. A light that was not of earth shone around and through him and his magnificient eyes were filled with an intense power that seemed to pierce his earthly companion through. Constantia had never before seen such a being as this but he was not afraid, a deep emotion of awe mingled with wonderment stole through his frame but with this no element of fear was felt.   For a moment the two entities were motionless, each gazing into the eyes of the other in profound silence, but presently the visitor from the skies spoke, his voice seemed like the soft strains of far away music yet penetrating and distinct.

Translated into the modern tongue, the speech of the celestial was as follows: "Oh Constantia, child of a favored race.   Thou hast been selected to guide and to represent a great and important nation.   Long ages before thy mortal birth thou wert a being of the planetary spheres; that earth might have a fitting soul to help to conduct its affairs and to develope its powers thou wert selected by wisdom guides to descend upon this mortal plane, and to pass an experience upon Atlantis.   Stancia, thy consort wert thy companion in the planetary realms, and she too came to earth to be they helper and thy mate.   For knowest thou that Stancia is a great seeress and an advanced soul.   Only once hath thy wife conceived and bore a child.   This child is thy infant son Fontis.   He is a child of the Gods.   Fontis will grow and thrive, he will become a great teacher and a leader of the people.   From his loins will descend a race of seers and teachers who will be high priests and leaders in the closing years of Atlantis.

"Knowest thou Constantia, that the continent that thee and thy people do live upon is doomed.   Its course

will run three centuries when it will be swept into the
sea, and all the world in ages yet to dawn will mourn its
fate.

"But so is it written in the history of earth. Planetary
law and action hath decreed the doom of Atlantis.

"Long ere its fate shall overtake thy continent, thou
and thine will have joined the planetary beings of the
spheres. Thy descendents, the children of thy Fontis'
children Cabul and Iberna, the grand workers who are
brothers even now in planetary life will bear a part in the
closing scene of Atlantian life, and will go down with the
continent to its doom. So hath it been decreed by the
law of Universal fate.

"Constantia, take this scroll, bear it to thy home.
Open not until thou join thy wife. With thy Stancia
read thou the contents of this scroll, above the sleeping
form of thy infant son. Mark its contents well, nor let
them pass from mind, for thou art chosen as a messenger
to thy people. A messenger of truth from the everlasting heavens.

"Thou doth wonder oh Constantia who thy visitant
from the planets may be. Know thou oh son of earth,
that he who now speaketh with thee, is a dweller in the
planetary spheres of Mars, a worker of wonders and of
signs, one who was thy sire ere the planet earth sprang
into being and was sent forth whirling in space. Thou
art my Son."

As the strange visitor who was still seated upon the
ground uttered these words and declared his relationship
to the president of Atlantis—the latter felt a thrill of
emotion surging through his being, and there opened to

his sight a scene of surprising loveliness. A scene of some far off past, in which he, as a spiritual entity had figured. A scene, that in its matchless glory of sentient, active life in exalted planetary spheres of beauty and light cannot be described in mortal speech. For a few moments this vision held his sight then slowly vanished, but not before he had seen himself as he had been, magnificent as a planetary being, and knew that he should yet again become like unto the majestic being who had come to give him advice. For now it was the turn of Constantia to speak, to return thanks and recognition for what had been imparted to him not only at this interview but also at various times through the seership of Stancia. To pledge himself to rear his son Fontis in such a manner as could only bring out the noblest qualities of a future leader, and of a child of the skies who was destined to become the sire of lofty men. At the close of his speech Constantia became aware of another presence by his side and looking up he discovered that it was Stancia standing apparently in mid air and regarding him with smiling and approving eyes. But it was not Stancia in the mortal who had thus glided to his side, but Stancia in spirit who had left her mortal form asleep in her own private apartment and had wandered to join her mate.

At her approach the planetary spirit made salutation to which she responded, for in spirit she knew him well as one of the great and glorious guides who had to do with the affairs of earth, and who had previously been presented to her as one of her special guides, then the visitor passed the scroll in his hand to Constantia and as he did so he arose, made a gesture of farewell and sailed forth into the upper air vanishing in a moment as disappears a wreath of mist. But Constantia held the scroll in his hand, he also knew that it had been no myth or phantom that had attended him but a veritable human being visitant from another planet, as he turned towards Stancia he realized that she trembled, and that she was breathing these words to him:

"The body calls me, Fontis awakened and needs my care. I go, follow me at once." In a moment she too was gone leaving Constantia in the deepening twilight to make his way slowly and thoughtfully home.

An hour later Constantia and Stancia were seated beside the tiny crib of their infant son. The little one lay upon his bed in a semi slumber, his rounded limbs and parted lips making up a pretty picture of infant beauty and innocence.

Constantia unrolled the scroll in his hand, upon it were imprinted golden characters such as he understood. He began to read them aloud whilst Stancia listened with strained ears and bated breath, for this herald from the spheres had come to them in a most uncommon way.

It opened thus:

"To Constantia
    and to the people of Atlantis
        Greeting from the Planet Mars,
Greeting and love from Kindred on the Planet Mars.
Greeting and instruction from helpers on the planet Mars!

"To the people of Atlantis, Greeting!
Know ye that to Constantia and Stancia hath been born

a male child, even Fontis the offspring of planetary love, know ye that Fontis hath been sent to earth to open out great storehouses of wisdom from the Supernal realms, thus be it declared that for the period of twelve years Fontis must be isolated from the world and held only in charge by the high priests of all the councils. At the age of thirteen Fontis will appear to the world and then shall light and wisdom be given the nation through his speech. For twelve years will Fontis lead his people then shall Fontis and such young men as he shall select, twelve in number, set sail upon the sea to unknown lands. Fontis and his men shall find favor in the distant islands and they shall take to themselves mates and rear offspring and found homes in these far off lands. They shall teach the people of those lands, of Atlantis and give to them its traditions and legends with a prophecy of its doom, and these records shall be preserved that future ages may learn of them. When twelve more years shall pass, Fontis shall return to Atlantia the land of his birth, leaving his friends and they who were his former fellow voyagers behind him, and they shall travel in ships to divers countries and found homes and settle families there. But Fontis will become a new leader and teacher to his people. Then will Fontis become the sire of children who will in turn become great men, and these two children will each beget a Son who will be leaders and teachers of men and who in turn will beget great men who will be leaders of different districts when the Doom of Atlantis falls. These last shall be known as Cabul and Iberna, the great and good. All this is prophesied from the Planet Mars by the prohets who read the histories of worlds and who mark

the career of individuals. Thus is the prophecy made
that the people of Atlantis may hand it down to future
generations as a sign of the wisdom and the power of planetary souls.

XXVIII.

And all things came to pass in the history of Atlantis
that had been transcribed upon the scroll, which the planetary spirit had mysteriously borne to Constancia. And
Fontis the child grew and waxed strong in the goodness
and grace of youth. As Constantia the president of the
republic was at the time high priest of all the councils so
had he the power and privilege—even to the Planetary
instructions of keeping his child under his own control
and instructions for a period of twelve years during which
time the boy associated with no one save his parents, and
the spirit entities who came to instruct and bless him
from the exalted planes of immortal life. At the age of
thirteen, as had been declared, Fontis appeared before the
world as a teacher and guide to the people, he was attended
by one wise man of Etruria—travelling over the continent
preaching and blessing the people—for through his
agency many new and wonderful things were revealed.

At the expiration of his allotted twelve years in this
work Fontis with twelve selected young men embarked on
their ship, and set sail for a foreign shore, guided by
the instructions of the high intelligences who ruled his
life. Fontis made his way to a far distant port in the
Pacific Ocean, where the extreme western portion of
North America is now situated. There he and his

friends found a tract of land that was fertile, and was peopled by a race of bronzed tinted human beings of strange speech and of gentle manner. These people gave welcome to the strangers and entertained them in their huts or tents of bark and leaves.

The people of this new shore although kindly and intelligent were far inferior in point of skill and mechanical execution to the inhabitants of Atlantis, and the young Atlantians set to work to teach them many things which would enable them to add to their comfort and knowledge.

Here, Fontis and his friends lingered upon this Island, and others which they visited in the vicinity, building themselves pretty homes and taking to themselves wives from the daughters of the Chiefs of the tribes. These maidens were gentle and beautiful creatures of Nature who easily learned the tongue and the customs of their Atlantian mates, and soon came to intelligently converse upon the subjects that were presented to them. They were comely of face with long and beautiful black hair, large and lustrous eyes, finely carved features and of graceful figure. Of these wives, were born to their Atlantian mates offspring who were reared according to the customs of the Atlantians, and unto whom as to their native mothers— —was imparted the history of Atlantis as well as the prophecy of its doom.

Fontis had chosen for his bride the sweet and pretty daughter of the head Chief of the Nation, and she had gladly responded to his love. She was a sensitive creature of great psychical power and through her mediumship, her mother who had passed from earth before the coming of the Atlantians had predicted her union to a strange man and one who should come to her home from a far off land. And all that had been told did come to pass and Fontis married the beautiful child of Nature whom he adored and their home was an Ideal one. Fontis and Fontina—as he called his bride—opened a school of manual and intellectual training into which were taken all the young people of suitable age, and many of the older ones as well and many things of value were imparted there while all the people were growing in such knowledge and experience as they had never before.

But time passed and Fontis had been the husband of Fontina for ten years, when the aged sire of his spouse passed to the higher realm. At his death, the people wished to make Fontis Chief of the Island—but he would not have it so, declaring that his time among them would soon be out and that he and his family must in a year sail away to Atlantis his boyhood home.

Then Justa, one of the boon companions of Fontis and Justina his wife, were made rulers of the tribe and the people were well content.

Another year passed and Fontis made it known that he must not longer tarry away from Atlantian shores.

Then the vessel that for twelve years, had been utilized in the near waters of the Island, was refitted and made ready for Fontis and his beloved ones, and at last the hour of parting came, and the dear home and the dear people who had grown into the very life of Fontis and his family were soon left far behind. In due time the voyage of Fontis was made, and at length he arrived at the home of his earlier years.

But as is known he came not alone for although he had left behind him all who had twelve years before sailed from Atlantis by his side, he now returned to the beloved continent not only with a beautiful and gracious wife, but also with a charming little son who had been born upon ship board midway between the island birth place of his mother and the Atlantian natal place of his sire. Beside his wife and infant son, Fontis brought with him two of the native sons of the island and their wives who had refused to part with Fontina, and who elected to go where ever she should go.

As well may be believed there was rejoicing and great display made not only at Etruria but in various parts of the Continent over the return of Fontis.

His parents who were still living were waiting to welcome him with open arms, and all of his former friends stood ready to take him again to their hearts.

And so it came to pass that Fontis settled in a home at Etruria and did whatsoever the spirit commanded him to do, and in time he became the sire of another male child, and he devoted much time to the training of his children, and also in conducting the affairs of the people who had placed him in high position.

And as the time went on all things became fulfilled as had been foretold although as if a part of the prophesy would fail for to the descendents of Fontis offspring were not born until the male parent had attained to a great age, so that by the time Iberna and Cabul came to earth more than two hundred years had passed from the day when Constantia—who in the fullness of time had passed on to Planetary Spheres—had received the scroll from the mystic visitor in the grove of Etruria.

But in time all things were fulfilled, and in the last Century of Atlantian History, Iberna and Cabul respective sons of twin brothers, were born, and these two children grew to be men and leaders of men, and each was given office of high priests over seperate districts which office and leadership were still theirs when the awful calamity occurred that swept a Nation and a continent from the face of the earth, and all that had been prophesied and commanded had come to pass.

But on the far off island that had once been the home of Fontis and Fontina the Natives lived and thrived, and the descendents of the twelve who had taken up their homes there on, grew powerful and strong and from them was born sturdy stock that also reared offspring, some of whom sailed away to unknown seas and made homes for themselves in foreign lands.

And some of these found the land of Asia and established homes there and became the progenitors of other races and tribes, and although terrible storms sometimes lashed the earth in one quarter or another and tidal waves swept in fury over the land here and there, and convulsions of Nature in one place or another seemed to threaten the planet with the crack of doom, yet in one part or another some human creatures, survived the shock and storm, escaping from one part to another of the world when the fury of the elements threatened to sweep them down, thus from these remnants of tribes and races, were preserved specimens of the old Atlantians—and other—people, although in a degenerated state, for being at the mercy of the wind and wave, and beaten about from place to place they could not preserve the glory and dignity of their ancestry nor the prestige of great presences from beyond. Struggling and coping with the conditions of poverty, famine, tempest and danger of various kinds, these people did well to keep their footing on the Earth, and to preserve the legends and the traditions of Atlantis and of the Pacific Isles.

But all had been fulfilled and the glory of Atlantis as a grand and progressive world and race, has departed from the face of the earth. Yet the noble old Atlantians live in higher realms. They are now Ancient spirits— some of them even consorting with Gods and Tutelary beings. Some of them achieving the marvelous works of Gods. They are living, grand, and conscious entities, who can engage in making worlds and in bringing the affairs of nations into line with the progress of the ages.

They still live, not as mummies nor as fossils of a decayed past but as active entities whose power is beyond the scope of mortal expression, and whose glory cannot be depicted by human tongue.

XXIX.

In the fullness of a ripened age, Constantia beloved of all the Atlantians passed from earth. Stancia had likewise departed, but although it might be said that she had died, as her physical form had yielded up it's spirit, yet one could not make her aged consort Constantia believe that she had departed from him. For there was scarcely an hour at a time when the rejuvenated and beautiful

Stancia was not by his side. Nor was she invisible to
him or to her people, for the inner sight of the aged man
had become so keen, he could behold his wife at any
time, and as for Fontis and his family each of them were
blest with spiritual vision and were able to behold the
beautiful Stancia who had become their guardian and
guide. For a few years after her translation Stancia continued to linger by the side of Constantia, and then she
informed him that for half a year she should absent herself after which she would return to him and conduct
him to his spirit home, for at the expiration of that period
his heavenly abode would be ready for him. It would
be necessary, she said, for her to be absent for a while as
he was living in part on the magnetism which she imparted to him, and the guides of the wisdom spheres had
admonished her that in order for her to draw him to the
spirit world, she must leave him long enough for the
magnetic forces that he had already imbibed to lose their
energy that his hold on earth might become lessened.

Then did Constantia learn that his time to pass from
earth had come. But there was no fear or trembling in
his heart, for well did he understand that the passage
from earth to immortality is but a step, and that no spirit
can lose his hold on life.

There was no fear of death in the far off period of Atlantian history. No tale of endless punishment or
misery had even been made known to the people. No
thought of an angry or vindictive God had ever disturbed
the human mind.

Legends and traditions of the past only told of a
happy, peaceful, and intelligent humanity that had risen

to worlds of beauty and gardens of delight from which they scattered blessings down upon the new born generations of earth.

And the Atlantians knew they had nothing to fear. They knew that death was but the deliverer to open wide the door of escape from narrow conditions to broader environments, they knew that omnipotent love must be superior to finite love, and they understood that earthly love would not see its offspring fall.

Therefore they were not afraid. Inspired teachers had taught them of the higher life, and had shown them how to advance in high culture and spiritual power. Thus the beauty and the purposes of Life were shown to them, and they looked forward to the hour of their deliverance with satisfaction and joy, preparing for it as one in these days prepares for the wedding hour or for the advent of some joyful occasion. At the same time each Atlantian was content to wait his summons to the spirit world, and to do his works on earth with fidelity and skill. He knew that it was wise to live as long as possible and to take the best care of the body that the spirit might express itself with intelligence and power, therefore, while he anticipated and knew the larger life that awaited him in the Spheres, yet, he was prepared to linger on earth and reap his experiences until the summons came for him to go.

For three months after Stancia's departure, Constantia continued to be busy with his affairs after which he gradually withdrew from the world, and passed his days in silent contemplation or in serene soul communion with the exalted ones of the heavenly spheres. Although aged—as to years—the form of this patriarch was as erect as in his earlier years, the bloom upon his cheek and the sparkle of his eye were also bright, and while the whole body wore an air of vigorous strength that was rather the embodiment of health and of life than of waste or death.

But as the hours of his voluntary seclusion wore on a marked change came over the man. His form became gradually shrunken, the bloom left cheek and lip, the entire framework seemed to be growing smaller and only the light of the deep dark eyes remained—a light that was more of the inner than of the physical man.

All the world knew that Constantia was about to leave them—but there was no mourning or sign of loss or woe. Every one rejoiced at the good fortune of their beloved leader, who had won the distinction and the honors he was about to receive, for who could dream of mourning for one who had passed his initiative degree and grade and was about to be promoted to a higher state?

While there was no royal station for any one upon Atlantis, as the Nation was a republic in Government and in sentiment yet special colors were used on special occassions to mark the expected transition of one in high position or of great dignity of character, and thus, the regal color, purple, was selected as the draperies and festoon with which the great Temple of Etruria was to be decorated at the approaching ascension of Constantia. For three weeks before his departure the grand old man daily received visits from the high officers of state who came to him from the various districts of the Continent, who came to hold audience with him concerning his going

to another world, and to give him messages to one
and another of their people in the spirit world whom
he would meet in his journey through the spheres.
During these weeks the Temple and the various
public buildings of Etruria as also was every Temple on
the Continent, were hung with draperies of the richest
purple tint and of soft and silken hue.   But if the drapings were of regal color they were not of woe, for they
were simply the sign and token of a great souls passage
from death to eternal life.   Three weeks passed, and the
aged sire knew that his hour was nigh—for in the soft,
balmy watch of the night he had been awakened by a
gentle touch upon his hand, and he beheld Stancia beside
his couch.   She was more beautiful now than he had
ever seen her before.   Beautiful in all the radiance of a
Planetary Angel and as she smiled and beckoned to him,
she seemed like some enchantment that presently might
fade away.   But no, she lingered and obedient to her
gesture the patriarch arose and donned the snowy robe of
ascension that had been waiting for him these many days,
then seating himself by the window that faced the dawn
he waited for the end.   A celestial light shone on his
own visage, which was lighted up with newborn beauty
and power.   His form had dwindled away during the
three months past to seemingly but one half its former
size, for the guardian spirits who had been protecting
him had been drawing from its elements and forces from
which his spirit body was now composed, and that portion
of his being that would be left behind was but the purely
physical of which the spirit would have no part.

The sound of celestial music now rent the air filling

the dwelling with richest melody and rousing the sleeping inmates to the knowledge that the supreme moment had come. Presently Fontis and Fontina appeared at the door, and in a moment more they stood by the aged leader's side. But he was not there, only the clay cold form so slight and wan which they tenderly laid upon his couch, for painlessly and without a sigh Constantia had slipped from the mortal, and like a bird let loose was now soaring with Stancia by his side into the blue ether towards the planetary Zones.

The signal of release was given and instantly the purple drapings of the buildings were removed, and snowy hangings of richest lace were festooned in their place—for a soul had passed out of bondage into life, into freedom, and what so fitting to tell the tale, as chords of sweetest music, and hangings of purest white as emblems of peace and rest. And so sweet music rang forth, and all the city felt the harmony that heaven sends forth at the ascention of a lofty soul.

Then went the signal forth over all the land, and the public buildings of each district lost their purple hue and assumed the badge of snowy white, and for two weeks these lacy streamers fluttered in the gentle breeze.

For a week the body of Constantia was laid in the temple over which a mighty mystic service of the ancient order that had been founded through the mediumship of Stancia was held, after which it was consigned to the funeral pyre and burned upon its bed of fragrant flowers and spicy boughs.

These Ancients believed in cremation and they practiced it. They had no grave yards filled with lowly mounds, but the bodies of their dead were consumed by the fire, and their ashes strewn to the wind, for they argued that all that was potential for conscious, sentient, intelligent life had gone from them and that these ashes belonged to the earth.

Thus passed on a great and good man. Constantia who had for thousands of years moved in circles of light in the great Beyond, and whose power and strength have made many human souls blessed, for he has uplifted thousands from a condition of weakness and blight into one of strength and of new life.

Oh blessed soul Constantia, thou and thy immortal mate doth work in God like ways for the benefaction of thy fellow men!

XXX.

But the descendents of Constantia and of Stancia still lived after the decease of that nobly mated pair.

Fontis and Fontina with their beautiful children were now doing a work for humanity, as important and valuable in spiritual and in intellectual training as that of their elders had been. And years passed on bringing only the vigor and bloom of ripening powers and culture to the lofty souls, until the period approached in which Fontis at the advanced age of a century also ascended to the planetary spheres. As it was with his sire so was it with his noble soul, preparations were made by him to take a journey. Messages were given him by one and another of his people to bear to celestial beings whom he was sure to

meet in his travels through the spiritual spheres. And
like unto his sire too, did Fontis pass on without a tremor
or a sigh. Conscious of the presence of his guides and
relatives, beholding them in the clear light of the day that
streamed about him, he went into the spiritual realm, not
as one goes into an unknown land, but as one passed out
to a home and condition with which he is familiar and
that he is glad to reach. And there was rejoicing
through all the land, rejoicing because Fontis was so beloved. The people knew that he was a wise and good man
and that he had earned a high place and grand distinction
in the beautiful heavens and they rejoiced for him, and
there was the sounding of musical instruments, and sweet
singing of hallelujahs in every Temple of the land, for a
great soul had been freed from earth and ushered into the
kingdom of the blest.

Among the traditions of Atlantis was the following as
nearly as it can be translated.

"Once in the dim ages of the past two waterfowl rested
upon a broad green leaf in the midst of a marshy pool, all
around them was only miry clay in the midst of
which the pool reposed, and the fowl looked at each other
and wondered from whence they came, for they had been
brought there from a beautiful stream of life on the spirit plane, but here all was cold and dark, all at once a
great light shone around them, and as it fell upon the
rusty pool its waters became clear and limpid and its
depth glimmered white like burnished sand, and the fowl
were glad, and they flew from the broad green leaf and
skimmed the now sparkling waters of the pool, which
under the mysterious light extended until they became

the sparkling waves of a beautiful sea. And the light still shown, and the dark and miry banks of the sea turned green and brilliant with blooming flowers. And still the light shone on and the mated pair of fowl swam to the shore and made them a nest, and soon in that little nest an egg was laid. And the Mystic light shone on. It warmed the egg in the nest until its shell was pierced, and from which protruded a tiny head and the light shone on, and the little form in the egg outgrew its covering and it lay upon the flowery sward. And the light shone upon the tiny form that had outgrown the egg, and the form straightened and grew but its little body was not covered with the skin of a bird but with a soft and downy fur or hair, and the light shone on, and that which in the body of the waterfowl had been an incipient wing, in the body of the tiny thing became an incipient hand; and the light shone on. Then that which had grown from the egg raised itself in strength from the sward and it stood upon two feet, and stretched forth two hands, and its head was round and the face was that of intelligence and the body that of a man, and the light shone on. The waters of the blue sea sparkled and shone. The mated water fowl riding upon the crest of the wave came and looked upon the strange thing that had come from the egg, and when they saw the face of intelligence and the form of man covered with it's downy coat of fur, they each arose upon spreading wings and sailed aloft far up in the ether blue, to return no more. But the light shone on, and the man walked in its rays and grew strong, in the light of the mystic ray, he wandered on until he came to a great tree that spread its branches

low to the ground and amid them he found a resting place
and food, for the treee was covered with rich fruitage and
it was pleasant to the eye and sweet to the taste, food
and drink was the fruitage of the tree to man for its
juices were like clear waters that were as honeyed dew distilled from flowers, and in the fork of the tree where
branches met, he made him a couch where he slept and
dreamed of immortal powers and of spirits of the air, and
the mystic light shone on, and the days and nights rolled
a pace and still the man grew, and the fur coat began to
fall away, little by little its hair fell out—and the light
shone on, and in the mystic rays of the light the fur covering was replaced by a fine and snowy coat of skin that was
soft and beautiful to the sight, and again the man slept.
But his coat of hair was gone and he felt chilly in his suit
of skin. Then came to him in his sleep, while yet the light
shone on, a maiden fair as a summers dream, a maiden
out of the skies who was clad in gossimer robes of fleecy
blue, a maiden with yellow hair and eyes of violet hue, and
she bore over her shoulders a great garmet of crimson
beauty, and this she took from her back and wrapped it
around the sleeping man, and no more he shivered in his
dreams for it had brought him warmth. And the light
shone on, and in it's rays the maiden glowed as a beautiful
summer rose. She touched the sleeping man with her
dainty fingers and he awoke, gazing upon the wondrous
vision of delight the sleeping soul of the man sprang to
life, and he held out his arms to the maiden of his dreams.
And light shone on, and in it's mystic rays appeared two
majestic beings who were spirits of the planet, and behind
them sailed the water fowl that had sailed above the glist-

ening sea and had become birds of the upper air. And
the majestic beings smiled upon the waiting pair and blessed them. And in the mystic light the union of two souls
on earth was formed.

But the light shone on and the lovers descended from
the leafy tree and hand in hand they roamed along the
grassy sward. And he drew his crimson robe of love
around him and smiled upon his bride, and she in all her
fleecy daintiness of her blue garment of faith received his
smiles and was glad. And lo, the majestic beings attended their steps, and by and bye they came to a beautiful
dell surrounded by flowering and fruitful trees and in the
centre of this lovely spot a snow white tent was spread,
and they entered while the light shone on.

Then came the foundation of the happy home where
love and truth enwrapped the trusting souls, and he
shared his crimson mantle with his bride, and she took
from her robes a portion of the fleecy blue and girdled
him about and love and faith became one soul. And the
light shone on, while the majestic beings lingered at the
tent, and the soaring birds came and brooded above
them, still shone the mystic light, and while all the world
but this sacred spot that had sprung into fertile beauty
in its gentle rays, lay a waste of waters and a chaotic
mass of mud, the beautiful dell became the garden of
creative force and life.

Thus the lovers wedded and the planetary beings tarried
with them and by and bye there came to their delightful
haven a tiny soul, a babe with kindling eyes and skin as
white as milk and it flourished and grew for the light
was never quenched.

And the babe became a man and still the light shone
on, and he learned many things, how to spin a silken
thread and to weave a shining web, and many other
things he learned, and by his skill the tent became a
house, and by and bye a temple was reared of beauty and
of power.   And the light shone and in its magnetic rays
a maiden walked, a maiden fair to see who had journied
out of the upper kingdoms of the blest, and he found fav-
or in her sight and he became her mate and they were
wed, and happy offspring graced their home, and the light
never waned, and all the glory of it fell upon their home.

But the majestic Beings had disappeared, the birds of
the air had vanished—the world was rocking in the throes
of pain .

Yet the light shone on, and the peaceful valley held its
own.   Time passed, and yet another race appeared,
coming out of shadowy mists of upper air—and they amalgamated with the first and homes were built and a city
reared its walls where chaos once had reigned.

And through the years the mystic light shone on, until
the world grew out of turmoil, and ceased to rock and
fume.

Thus nations came and great Atlantis founded by majestic beings from celestial life appeared to crown them all.''

Such was one of the traditions handed down to the
people of Atlantis, and when ever a great man died, a
pair of water fowl were borne far out upon a lofty mountain and let loose to sail away into the ether blue, as the
herald of the coming of a mighty soul.   So had it been
with Constantia, as also was it done at the death of Fontis, and too whenever the child of a great family was born,

crimson and blue flowers were hung upon the Altars and shrines of all the temples in honor of the robes of Love and Faith that had clothed the first pair and consecrated their nuptial ties.—So had it been at the birth of Fontis as it had been at the birth of his succeeding heirs.

XXXI.

The life and progress of the people of Atlantis continued to advance through the passing years, and in time, Cabul and Iberna, descendants of the great Constantia and the gifted Stancia appeared upon the scene. These two were sons of two brothers, and a strong magnetic cord held them in fraternal fellowship as if they were brothers, born of one flesh, as these youths advanced in learning and in skill the people realized that it was all in accord with the prophesy of the Planetary spirit that had been traced upon the scroll presented to Constantia in the grove of Etruria by his mysterious visitor. Therefore were they given special attention and training by the high priests of the land, but in a few years they had outstripped the learning of the Sages and Philosophers and were themselves well qualified to teach and lead even the most advanced among them.

And it was known by the people of the country that Cabul and Iberna were of the planetary race, and that they were high intelligences who had spent ages in the exalted spheres of planetary life, but who had elected to come upon earth as teachers and leaders of the race. And it was also known to the Philosophers and Sages that

these kin would be upon the Continent when its doom
would fall.

But throughout the various districts the people but
feebly realized that the end would come within the Century, for although as a rule they believed all the prophesies
made to them from Celestial Zones, yet they seemed to
feel that perhaps this final one would be unfulfilled; or
at least but that a portion of their beautiful land would
be submerged.

But a few remembered the prophesy and because of it
they sailed away to foreign shores there to establish new
homes and to perputate their traditions—and still others
remembered but they believed it was to be their fate to
go down with the Continent and thus to gain their en-
trance into the Celestial realms.   As the years advanced
Iberna and Cabul were appointed Governors—and High
Priests—over large districts, that of Cabul was an inland
territory not far from the central seat of Government
while that of Iberna was at the ancient city of Hermenia
that faced the sea.   At their respective districts these
leaders performed their duties and filled their offices with
true dignity, and with spiritual power, and each took
to himself a wife of the Ancient lineage and of high birth,
but while to Iberna was raised a son of intelectual vigor
and spiritual power.

To Cabul all offspring was denied.

It is not our purpose to dwell at length upon the lives
of these great Atlantians for we have given sufficient in
these papers to acquaint our readers with the trend of
work and experience which were a part of the lives of the
leaders and teachers of the land, and these last, did not

differ essentially from the life of their ancestors unless it were in even a greater manifestation of spiritual power.

Now be it known that in the traditions of the Atlantians it was held that no male personage could arise to the potency of Spiritual impulse and power sufficiently to gain guidance over a planetary sphere, or to become associates of the Tutelary God of any planet who had not rounded out at least a century of planet experience upon such a physical planet as he should happen to be born upon.

And indeed that might also be said of the female Companions unless they had spent a long period of service as Oracles for the divine wisdom of the spheres.

Therefore if the son of a high priest—who was also a reincarnated planetary being should happen to pass from earth life, below the century mark, it was expected that he would be reincarnated, and become again a dweller upon earth, for at least the term that had been cut from his former experience and labor and if he should remain to a longer period, that was counted in his favor on the Celestial side.

As we have said there was born to Iberna and to his beloved consort Irnie a son. A son, who according to the custom of the times remained with his parents at the temple of Humenia until his thirtieth year, receiving instruction and guidance from the sire whom he loved with the greatest of filial affection. At the age of thirteen he began to preach and to teach, going into the outlying towns and dispensing the truth that came to him from invisible realms. At the age of eighteen this youth was initiated into the Symbolical Order that had been founded by planetary beings through the agency of Stancia centuries before, and he soon became a master of high degree in the assembly, as his occult powers seemed to wonderfully increase after he had begun to learn the mystic signs and symbols of the ancient cult.

At the age of thirty one, the son was taken into the Temple and given many instructions by his sire, among the things that was told to him was the following:

"My son," said the aged Iberna, "the time hath come for thee to pass through this land, and to bear the light of thy spiritual counsels into many homes. Thou wilt meet with strange experiences, which cannot be made known to mortal man, and thou wilt be attended by wise intelligences from beyond. Thou wilt not return to Humenia for thy work will hold thee in distant places, and thou wilt behold the city of thy birth and the parents of thy life no more. Yet wilt thou arise on the day of thy death to the planes of spirit life in which thou shalt work and think. Take then, tender leave of thy mother for ere the year shall wane, her soul will wing its way to yonder stars. Nor wilt thou behold thy sire again, for ere thy face can be again turned toward Humenia, Atlantis will be swallowed up in ruin and all its souls be translated to another world. Nor is it best for thee to sail away and escape its doom, for thou wilt have work to do in spirit with many who shall enter the life of the immortals. All who are of sufficient soul growth and training my son shall pass into the higher spheres, or to planetary realms. Many there will be of earlier years like thyself who will not pass on—to them wilt thou be a teacher and an associate. Thou wilt labor with and for them until such time as they ascend to the Spirit Planet of this earth, or until they find reincarnation in a future age upon the earth. For some there are among them—and thou art of the number who will in time become Tutelary Gods of either Zones or Planets, and these must complete a century of experience and work upon earth.—Thus will they be reborn and live for years upon this terestrial globe.

"Thou knowest, my son, that a spirit continent is being formed from the emanations of this Atlantis and its people—that is, it is situated above this physical planet and that it cannot be disturbed by volcanic eruption or tidal wave of earth—that it has minute resemblance to this continent even to its seas and rivers its valleys and hills. This spirit counterpart is for the use of those who shall not round out their natural lives upon Atlantis, and they shall take up an abiding place there, found their schools, build their temples established their Homes and gain their experiences—those among them who are not to be the Gods of spheres will not have to be re-embodied on earth—but they may be so if they choose but when they have gained their experience can pass on to higher Zones or to the Spirit Planet.

Those who are to become gods and guides of Zones and Planets will dwell thereon as teachers and guides to the people, until this spirit Continent is no more. When they will dwell in planetary spheres until their time for reincarnation comes.

When all have graduated from the Spirit Continent that have gravitated there from Atlantia, and no more remain to enjoy its fruits and scenes, that spirit country will dissolve into a mist and its essential elements will be ab-

sorbed by the atmosphere of earth and again imparted to
regions of this planet that can appropriate them to new
forms and manifestations of life.

And all these things shall come to pass my son, for they
are revealed to me from the higher realms, and thou shalt
be reborn on earth and pass through experiences of much
import and live out thy life to the rounding of the century thou shalt not find in thy present incarnation.   So
be it.   All is well.

Now be it known to thee that thou shalt never be for-
saken for Iberna thy sire will be thy guide.   And he will
lead thee through strange places and thou wilt meet with
many things.

When thou goeth forth from this temple thou wilt journey as the spirit directeth thee, and thou wilt in time be
in the district of thy kinsman Cabul. Abide with him for
a time for though it be not made known at what hour,
the visitation of the fire and the flood shall come, yet will
it be well for thee to tarry with thy kinsman Cabul, that,
if the spirit willeth so, thou shalt be with him in labor
when the fate befalls Atlantis, gem of the sea."

And all these things were held fast in the heart of the
young man as he journied forth.   And it came to pass
that his mother Irnie soon afterwards soared to the
spirit world, and in about three years from that time, the
end came and Atlantis with her glory and her people was
swept into the sea.   But on the Spirit Continent a race
and a Nation dwelt and all the people grew strong and
happy until the lapse of thousands of years they ascended
to other realms, and dissolved the spirit continent and its
essentials were taken up by earth, entering another con-

tinent even that of the North American world as a potential force and constructive power.

XXXII.

The region of the North American Continent of earth is
by no means a modern hemisphere, for that portion of
the globe has existed for many thousands of years, and its
lands have been roamed over and built upon by human
beings in remote ages even as these entities have sailed
upon its seas, and skiffed along its shining streams.

People of strong and positive temperament dwelt upon
the shores of the Atlantic ocean and roamed over mountains and plains even to the far off western slope.   They
were of different races and tribes, some of whom were
more warlike than others but all were simple in manner
and of childlike character in one or another direction.

It is not for us to state the age of the planet earth, nor
for how many thousands of years it has served as the
dwelling place of man since the advent of the cave dwellers
or the evolution of the human type from the higher forms
and reincarnated life principles of animal life.   Nor can
we extend still further back in the history of the world to
that period of time in the remote past when a body of
land and all its beautiful essentials for a finely organized
type of human life to enjoy, were materialized by planetary spirits, out of the heart of the seething boiling planet
that would not for ages be capable of its own forces and
resources of sustaining life; and when for the good of
earth and its future ages glorious human entities from the

spheres of advanced planets materialized for themselves forms and became dwellers upon the enchanted lands that the earth might receive inspirations of magnetic force from them and thus be assisted in its evolutionary work.

But all these things had been, ages before the growth of Atlantis as a continent and ages before either the Pacific or the Atlantic oceans rolled from shore to shore.

But as Atlantis had arisen and thrived, so had other continents and innumerable islands grown into being and become a part of the world, and in the remote ages of its history the western hemisphere was made up of one large body of land and of a large number of islands around which the ocean waves lapped and surged with ceaseless energy and sound. But the people of this western clime could not at that period lay claim to any special kinship with the planetary spirits nor with any of the southern and eastern climes who were reincarnated souls, for these people of the west and north were the races that had evolved from the lower forms of human life, and who were in every respect the products of that wonderful development process of energy and potential force which determined the "Descent of Man," yet is this a misnomer nor does it properly designate the true dignity and aspect of that grand evolutionary work for although it signifies that man descended from an ancestry that was rooted in the animal kingdom, yet it does not sufficiently emphasize the fact that this very descent became in its grandeur and beauty a glorious ascent for man, an ascent over grander and grander heights of growth and progress, of developement and of cultivation until from the lowliest conditions of the earth man has gained almost the apex of physical and spiritual growth.

At the period in the worlds history when Atlantis was at its height of glory, that part of the world where the North American continent is now situated, was peopled as we have said by simple races and tribes, who were not educated nor cultivated in any of the arts and sciences. But as we have before stated, individuals now and then started from the continent of Atlantis and sailed to foreign ports, taking up homes snd establishing their customs there—thus did portions of their islands and continents become somewhat like unto the Atlantian region absorbing magnetic force of its people through the intermarriage of numbers of its young men with the daughters of various races and tribes. We have mentioned the fact of a Spirit Continent being formed above the physical Atlantis and that this spirit world had been designed and created by planetary spirits from the magnetic and electric forces and elements of the physical continent and its people, that it might serve as a home and school to the people of Atlantis from the earliest years of age to the octogenarians and elders, for it had been decreed that not until each had rounded out a century of experience in contact with earth, could they be fitted to ascend to the higher spheres.

In time the spirit continent had done its work. Its people had gravitated to one portion of the universe or to another, some to the spirit planet, others to the sixth or seventh sphere of progress, others to planetary worlds or spheres in the solar system and others to various portions of the earth, either to be reimbodied as mortals, or to serve as guardian spirits as to the tribes or nations or individuals of the planet, to whom they could impart their magnetism and inspiration.

At the time of dissolution of the spirit continent, its forces were taken up by various ancient and planetary spirits selected for that purpose and borne to the regions of the northern islands and continents upon which dwelt certain tribes in whom were mingled the elements of the Atlantian race with those of other nations. And these forces and elements from the spirit continent were absorbed by the people of earth and they brought a stimulating power that had its affect in producing new types of being—or variations in being—that gave a new impetus to the spiritual nature of man, and the ages rolled on. Many young men from time to time wandered from the islands and continents of the western hemisphere to far portions of the globe founding homes in various quarters and amalgamating with the people of those other climes. Thus were the elements and the forces of the Atlantian race carried to all parts of the world, infusing each with a stronger and higher power for the blessing and progress of all men.

But in the life of the planet new conditions appeared producing differenciations of climates and of people. But the Aryian race which contained the largest quota of spiritual element and force and which was largely influenced and guided by the planetary spirits that had held old Atlantis in their charge, made highest progress in its advance and planted its people and status over all the world. Throughout Europe and Asia they spread, and from the Ancient waters of the Nile to murmuring streams of that portion of the globe that embraces the continent of Europe they dwelt, varying their speech and customs according to the differing conditions and environments with which they came in touch. But in this paper we have more to do at present with the primitive people of America than with those of other shores, for we wish to emphasize the fact that this portion of the globe received the essential elements and magnetic forces of the spirit continent that had disappeared and that these forces and elements impregnated the atmosphere with a new vitality that of itself brought a quickening force to the very life of man. And this new power produced within the races and tribes the spirit of freedom which became a part of their natures and which lifted them above the conditions and the elements of bondage.

Thus were the Atlantian spirits especially attracted to this continent and many of them became reincarnated upon it taking up new forms and experiences in physical life not only for their own growth in soul power, but in order to teach and to assist the people in their onward march toward a higher spiritual cult. Those of the Atlantians who did not become reimbodied on earth, but who elected to remain in contact with its inhabitants were such as Cabul, Iberna, and their class who came as guides and helpers not only to individuals but to whole communities and it was through the influence and labors of such high intelligences that the modern dispensation of spiritualism has been brought to the world. Under their influence psychics have been developed in this modern age and the glory of spirit worlds and works have been revealed not only to a favored few of the church and school, but to millions of the common people who have received the revelation with thankful hearts and have felt their souls quickened under its illumination to a nobler and diviner impulse and effort for the highest god.

For thousands of years these Ancient spirits have worked for the good of man on earth, while that portion of them who remained on—or in charge of—the spirit continent of Atlantis till it disappeared directed their attention more principally to that portion of the globe—and its climatic and human developement—that embraces North America, others of the planetary beings and Ancient intelligences labored in conjunction with the Tutelary God of earth over other quarters of the earth.

Therefore hosts of them were employed in the stimulating of human and planetary forces in the direction of Asia, others in Africa, others still in Europe, each company or host according to the age and dispensation doing its appointed work.

Thus in every age of the world's history mortals have at one time and another and in one manner or phase and another come into communion with the messengers and Angels of the spheres, and these ancient visitors have exercised an influence upon the affairs of men through all the cycles of the world.

Thus, Abraham sitting in the opening of his tent received the glorified guests who came to him in the guise of men but with the light of heaven, and they gave instructions to him and the patriarch entertained them for they had come from far even from the higher spheres of the celestial world.

XXXIII.

Various have been the lands, nations and races of the earth over which the Ancient Spirits of Atlantis and other prehistoric countries have held an influence, and with whom they and their forces have amalgamated for good works. Among those who came directly in touch with many Ancient Spirits who had once dwelt on earth and with others from planetary realms in space were the original Aztecs, for before the birth of this people as a race the way was prepared for them and influences brought to bear upon the country in which they would establish a Nation and a home, by planetary beings and Tutelary intelligences who knew that the Aztec race would display a grandeur of art and skill and of native good sence and intellect that would prove of the highest blessing to them as a race not only on earth, but in the spirit world.

From the earliest period of Aztec history to the very extinction of its people as a distinct and unadulterated race, there glowed a glorious line of ancestral power and dignity which came from the influence of high planetary beings, which gave a lofty and even exalted air to the entire race. The people were a gentle, skilled, intelligent folk, of perfect grace in stature well formed and of handsome feature, whatever of aristocracy there is in the natural and dignified being of a true manly and womenly humanity the Aztecs had, their knowledge of the sciences of Astronomy, Astrology, Chemistry, and kindred lines, was profound. They understood the law of mathematics to a wondrous degree, and geometrical studies were a part of the mental curriculum of the students who came to the temples and to the priests or teachers for instruction.

The teachers and oracles were instructors alike in religion, in science and in mental branches of the more ordinary class.

The religion of the Aztecs was a simple one. They loved the sunshine and gave this luminary homage, but beyond all signs and symbols they recognized a supreme source of all power which they worshipped as the creator of all—and in their tuneful chants and praises at early morn and at dewy eve, they rendered tribute to the Infinite and Eternal Soul. The Aztecs were a peaceable nation, they had no desire to prey upon their fellow beings, but if they were molested by tribes from outlying districts and other shores they could defend themselves with courage and zeal. Their time was given up to the pursuance of peaceful labors and arts. Their young men and maidens were all trained in a knowledge of industry and also in the artistic line of cultivation. Many of them were fine musicians, skillful artists, gifted sculptors, hammerers of gold and workers of precious metals and stones. Beautiful pictorial scenes in art and of tapestry and other stuffs were wrought with patience and genius. The men and women were of like standing in the community occupying equal ground of voice and will in all the important affairs of life.

The Aztec nation had its governors and rulers who

were of noble lineage and of wise judgement. Its councils were composed of equal numbers of men and women each of whom were expected to take part in the deliberations and to express themselves briefly and intelligently on the affairs of the day.

The temples and many other structures of the Aztecs were rich in architecture and in works of art. The engravers tool and skill were employed upon these edifices to the greatest of advantage, and each public building stood grandly forth as a monument to art, learning, and to industry, the pride of a nation and the credit of a race.

In the land of Mexico grew beautiful flowers and luscious fruits. The people ate the fruits and inhaled the fragrance of the flowers and grew more beautiful because of them.

During the progress of the race from a lower to the higher state of culture and honor, and grandeur in all that went to make up the glory and beauty of its nation and its life, the type of civilization that was evolved would compare favorably with the higher grades of intellectual and of spiritual civilization of the present era.

Among the Aztec people in the middle portion of its history several wise and advanced spirits from the spheres beyond elected to take up a mortal birth within its precincts and within half a century of time, these intelligences came again into an earthly existence. These were of both sexes, and in time they became clear seers and wise oracles who were set apart by the people as proper beings for the guidance and instruction even of the high priests and leaders of the Temples and of the nation.

Nor were they mistaken in their views for from the teachings and councils of these reembodied souls the country
took on a new power and prosperity that carried it on to
yet a higher mark of culture and of success in all departments of thought and of achievment for which it was
adapted and designed.

In the higher realms of the spirit zones dwelt in the
purity and peace of her celestial gardens a lovely and radiant spirit who for thousands of years had served the
human family upon one and another sphere, as an instructor and oracle.   Zaida the beautiful had passed on to the
seventh zone and it was her privilege long since to arise
to higher worlds if she so desired.   But as yet she felt
that her work and mission belonged to the people of the
lower spheres and thus she labored among them imparting a divine light and uplifting magnetism to all who she
reached, for all these ages Zaida had been without the
companionship and cooperation of a soul mate serving as
the instrument of highly advanced beings in her labors of
counsel and of ministration to other lives.   All around
her were numbers of exalted beings working in pairs, together, for the blessing of humanity but she as a special
oracle of the spheres had pursued her way without the
union that bringeth two souls into one sphere of harmony
and labor.

But now it became borne in upon Zaida that she had
ministered to the people as teacher and oracle sufficiently
long, that she should now have a change of operation and
of environment and that her experience must reach out
into conjugal relations with her other self whom she felt
must be awaiting her coming somewhere.   Then the

spirit entered into communication with the high forces of the soul with drawing into retirement she felt herself uplifted to the heavens far beyond the encircling spirit spheres of earth.

With all their beauty they paled into dullness beside the grandeur of this exalted and magnificent place, not words can describe, no pen depict the splendor of that supernal state into which Zaida passed, nor the significance and grandeur of the intelligence imparted to her by its lofty souls.

But in that hour of interior research and exaltation she learned that her own mate or counterpart dwelt as a human entity—a reembodied planetary being—upon the planet earth in the country of Mexico as Aztec adept in the mystic lore of the ages and spiritual cult of the times. That to meet and to become united to him she must take up her abode by his side, attend him in his works, influence and guide him, and through his seership make her presence and her relationship known to him, and thus Zaida turned from the glories of the higher realms to which she was entitled to promotion, and came to earth to the land of the Aztecs—and there in a temple of a beautiful city she found her mate even as had been shown to her, a teacher and a guide to his people. And her soul went out to him and she loved him with exceeding great affection. Then did she show herself to him and he became enraptured of her, for he too had passed the ages without a mate. And his soul went out to her and they blended together as one Angel, and he was never alone, and it came to pass that many who came to the temple to learn of the master beheld Zaida standing beside him or working in his light and they learned to recognize her beauty and to look for her. And under the power of this double ministration the glory of the temple increased and the interest of the people were enhanced an hundred fold.

But the day came when a foreign foe invaded the place and though the people of the city fought long and well, they were overthrown and many of them were slain, while others were bound hand and foot and taken into captivity. And those who were thus conveyed far away were released from bondage and their lives were spared on condition of their taking companions of the tribe of their captors and settling down among them. And they did so, and from this union offspring were raised who bore the traits of the Aztec mingled with those of another race, and a new type of humanity was evolved.

But the lover of Zaida was not spared at the siege of the city, for the temple was razed to the ground and its priests and teachers were put to death, but when the master was seized a great light shone around him and Zaida was seen standing in front of her beloved and shielding him with her breast.

For a moment the foemen were blinded with the glare but only for a moment, and believing this to be some mortal woman they rushed upon her to tear her from her hold. But as they did so a blinding flash rent the temple, and burying all, who had not escaped, in its ruins..

Those of the priests who had not personally been put to death were burried in the ruins of the falling pile. But these were only killed in body the spirit entities lived soaring to worlds of delight and beauty on high, and

among them were Zaida and her mate, and they were
united in all the sweet and holy affections of the soul,
and for a long period they dwelt in the wisdom of the
seventh encircling space, but at length they felt that if
they were ever to be guides of a world or even of a zone
they were to return to earth to labor for human emanci-
pation from error and sin.

XXXIV.

Now in the evolutions of the cycles of time Zaida and
her mate had grown into close relationship so that where
ever one might be the other would feel the pulsations and
know the magnetic vibrations of the other even though
worlds might roll between them. And it came to pass
that planetary spirits who loved them well, began to prepare conditions by which each might be born on the earth
and become a mortal worker among mortals. But it
happened that the conditions for the birth of Zaida was
perfected earlier by some years than were those for
Harma her mate who as an Aztec had lived for three
score years amid the glowing scenes of Mexico. And
thus Zaida took upon herself mortal birth being born of
intelligent parents in an humble home. And the child
grew and thrived, early showing signs of wisdom beyond
her years. And strong psychical powers were developed
in her organism and she became a seer and oracle to many
of the people who were uplifted and taught by her ministrations and her instructions.

And afar off the conditions were ripening for the mor-

tal birth of Harma and in time he too assumed the physical form and was reared in a strange country and among
strange but good and honest people.   Now be it understood that both these souls were in charge of high planetary beings who guarded and guided them and in time circumstances were outwrought that brought them together,
and after much trial and discipline purposely allowed by
their kindred, to affect them that they might prove the
quality of their own Atlantian metal and character—they
were united, and pursuing together a humanitarian work
for the ignorant and suffering that peopled the modern
age, they wrought out the conditions which were essential
to bring them still closer into rapport with the potential
forces of the Universe, and to quicken their grasp and
knowledges of superior things.

During their earth life these two knew but little of the
great important lives or career they had formerly led, for
despite their psychical powers and work they but rarely
caught glimpses of the past and its achievements, yet frequently in the night when slumber fastened upon their
outward senses they would soar in spirit to celestial worlds
and commune with their own people who were workers
there.

From the time of the Atlantians to the age of the nineteenth century of the Christian era, earth had passed
through many evolutions in its progress and unfoldment.
To a supernal observer watching the convulsive movements
of nature, and witnessing the struggles and trials of the
planet as well as of those races.

Nations and individuals who were fortunate—or unfortunate enough to be born upon it—it might seem as if re-

trogression instead of progression of the world and its people was the rule.

But to the advanced and enligtened spirits who knew that a world as well as an individual must pass through periods of darkness as well as light, sorrow as well as joy, storm as well as calm, these exhibitions of tumult and of shadow were only the inevitable result of the operation of the law of evolution and of involution which must work in order for the best results.

But earth had seen its shadow periods as well as its pleasant ones. Nations had risen and gone to decay.—Races had come into existence and vanished from the face of the earth. Continents and countries had appeared, flourished and disappeared, Dynasties had swayed with mighty rule and gone down into oblivion.

Yet earth rolled on its course and the disorder brought forth order, experience ripened into human wisdom, and chaotic uncertainty and unrest becoming illuminated with the light of the ages, blossomed into intelligence and skill.

Thus the ages and centuries rolled on—thus the seasons came and went. Tribes and races made thir advent and established their lands and homes, and the tide of advancing civilization rolled on. The christian era appeared upon the world like a sweet and subtile strain of music bringing forth revealments of a long lost and forgotten faith in the personal immortality of the soul. And like a strain of purest melody the religion of Christ stole from the soul of the humble Nazarene into the hearts of humanity. But the world had been defiled since the times of the Ancients. It had been corrupted

by false ideas and erroneous conceptions of Deity.   One
spirit entity who had for a time served as the Tutelary
to the lower sphere had become a fallen Angel.   And his
counsels and statements had been accepted as from the
Almighty.   The belief in a personal devil had also crept
into the religous notions of the ages, and altogether wrong
ideas of the universe prevailed at the opening of the christian era.   But the Nazarene taught and labored for the
common good.   He was a high personage in the spheres
who had elected to enter the earthly state as missionary
and exemplar of right living and of love, special conditions had been prepared for his coming and although he
was born of human parents like unto the human family at
large, yet he was specially anointed for his work by the
high Angels who gave him power, while for months before his birth his mother was overshadowed by the divine
influence of the higher spheres.

And the Nazarene came to earth—he lived His life and
performed His works passing into the glory of the heavens
to become a worker again for humanity.

And the Christian era crept on, but the folly and the
bigotry the selfishness and the authority of men instilled
into it the falsities and the glaring imperfections of their
own lives and purposes, and instead of remaining a religion of love—a religion of the common brotherhood of
man, a religion of faith in immortality for every human
soul, it became a religion of blood, of Idolatry.   Thus
the centuries moved on—many of the Ancients had again
and again been reincarnated to bring a special spiritual
influence directly into the world—many others remained
in the spheres as guides and helpers to mortals.   Wars

and famines, earthquakes, disasters and woes innumerable afflicted the human race. Suffering, disease and violent death attacked many individuals, poverty, crime and despair stalked abroad, yet, according to the worlds history the planet and its people were rolling on through a cycle of experience that would eventually bring a return of all the glory, grandeur, intelligence and wisdom of the past, together with greater potencies and powers for the production of a grander humanity upon earth.

And the centuries rolled on, beautiful nations and kingly races disappeared or became so amalgamated with other tribes and nations as to have their distinct identity swallowed up in the processes of evolution.

The Aztec nation like many others had lived its life of power, pride and beauty through centuries of time, but it too had become only as a name or as a tradition to the people of the new generation and dispensation.

We are not writing the history of earth nor of its lands and nations, therefore we do not follow the course of generations nor of countries in consecutive line. We have only given fragmentary hints of the lives of human beings and of certain portions of the globe, knowing full well that many nations and many people including Africa and various parts of Asia and of Europe have not been mentioned, even though they have all had their ancient history, as well as their prehistoric experience, and have held an important place in the earth's development. But as we have before said, it would be impossible to follow the line of unfoldment age by age and of planetary or human growth century by century, though such a history of prehistoric, ancient and modern developement and experience may in time be given the world as was never dreamed of by mortal minds.

We have written of the progress of the centuries of the Christian era, and of the sorrows and darkness that assailed humanity, and which continued to afflict the race in this modern age, and now we come to the nineteenth century.

A century that had been prepared for and looked forward to by the planetary and ancient Spirits who were in touch with this world. And the electrical forces of the universe were brought to bear upon certain brains that were quickened under their influence, and new wonders or rather old wonders of ages ago were revealed to man, and by its potency the possibilities of electrical force as applied to modern mechanics and machinery were demonstrated, and minds of earth, stimulated by the influence of the ancients were induced to think along lines of practical experiment and demonstration, and many things were made clear.

But the nineteenth century was only the vestibule or passage way into a new era of wondrous light and force, an era which will open more directly in the year two thousand, in which the so called Christian era will be merged in the humanitarian epoch of an unfolding race and generation of men.

An era in which the electrical forces of the universe will be employed in such utilitarian ways for the comfort and convenience of man, that the world will be an abode of luxury such as only the dreamer has dared to dream.

But the nineteenth century came, came with its potency and power of great unfoldments, and the ancients had pre-

pared the way. About midway in its reign conditions
seemed ripe for the advent of spiritual communication,
that knowledge of the after life might be conveyed to the
people of the modern age and America with its mixture
of tribes and nationalities its various elements and conditions, was chosen for the scene.

XXXV.

One of the special and important points that we aim to
emphasize is, that the planet earth has not been entirely
peopled by the operation of one law—or rather simply by
one operation of law.

The law of causation has produced every type of existence upon this or any other planet, and the law of attraction has determined the operation of the law of causation and its differentiations.   All things are governed by
law.

But neither special causation, evolution nor reincarnation cover the origin and genesis of human life on earth
and yet, each of these special branches of the law of cause
and effect is an important factor in the production of an
intelligent and progressive humanity.

In the earliest ages of the world planetary spirits elected to materialize a portion of the earth—or rather to
beautify it and by their electrical and magnetic forces and
emanations quicken its developement that it might more
speedily become a fitting habitation for man, than   it
could do were it simply subjected to the slow process of
planetary evolution—and when that portion of the earth

bloomed to their satisfaction these human entities materialized forms for themselves through which they could labor on earth, and build for themselves homes there, infusing into its atmosphere their own magnetic aura which went out in a stream of magnetism over all the earth assisting in its developement for future ages of man.

This materialization of human beings as inhabitants of earth may be called in a sense—Special Creation, and it is analagous to the Biblical Allegory of the first man whom God created out of the dust of the earth; while the garden of Eden of that story, is a symbol of the garden spot that the planetary beings created for the habitation of the first dwellers upon earth.

But as has before been told in this work, an evolutionary process of life was going on all the while and through the ages there appeared on the planet various stages of animal existence one type succeeding another until a higher form of animal activity, conciousness, and intelligence was produced, and when the bodies of these higher animals died the intelligent life principle remained in the atmosphere, not to be taken up in other animal bodies on earth but to be vitalized by the soul-germs of human entities that were floating in the atmosphere, and to be reproduced in this conjunction, as a higher type of being such as became known as the lower forms of human life.

And as the years rolled on these lower forms yielded up their life elements, and planetary action unfolding better physical bodies of brain power and thinking capacity, these life elements and soul-germs were again absorbed into matter, and herein was produced the potentiality of both evolution and of reimbodiment and again the planetary spirits desired to bless the world, and as the ages rolled on they concluded to lay aside the conditions of their life in the spheres and to be born on earth in mortal form to acquire earthly experience and discipline—and to become as teachers and helpers to their mortal brothers and sisters who had been evolved from the animal kingdom and who continued to exhibit the traits and properties of the brute creation—only in modified and humanized degree—from which they sprang. And thus the process of reincarnation extended its powers to these human entities and became a special law unto such as desired to avail themselves of its scope.

Now while the law of reincarnation works beneficently for the lower types and tribes of evolved human beings, and while it serves the purpose of those advanced human entities who pass under its provisions for particular purposes of growth, experience, or of achievement. Yet it is as unerring in its operations as is any other law in nature, and the exalted souls of the spheres who determine to enter earth as reimbodied beings know full well that for a period—be it long or short, they will have to lay aside their remembrance of former activities and associations and to merge all conciousness in that of the incarnation and identity that they assume when they take up the new form. But this they are willing to do, knowing that to the spirit and to eternity a century is as but a moment of time, and realizing that by merging all purposes and activities in the desire to bless humanity they may be able to help the race and earth itself onward to higher progress and a grander cult. Nor does any reincarnated, active intelligence remain ignorant of his ex-

periences before and after reembodiment when he passed
again into the higher spheres from which he came, and
should he elect to be successively reincarnated and form
parental or filial or fraternal attachments on earth, these
are well, for he can form no genuine attachments but
what are of the spirit, and in the larger planetary life of
the universe they are recognized as the links that bind
the soul to the great whole or fraternity of eternal brotherhood.

With the lowest types of human life that have evolved
from the animal kingdom it can be no question whether
they prefer to be reincarnated or not. With them it is
a necessity, all the laws of irresistible attractive force and
of gravitation draw them into contact with mortal life,
having evolved from the animal they at first exist in the
atmosphere as elementals, and are sooner or later drawn
in contact with the forms of lower human life. In these
forms they develope an activity and conciousness that
comes from the quickening force of the soul-germ they
have now absorbed, but they are not at first grade sufficiently vitalized with spiritual potency to be qualified for
immortal human existance, hence, on the death of the
body they remain as uncouth and dwarfed looking
beings in the atmosphere, until again absorbed into a
higher type of human organism in which the spiritual
gains greater activity and consciousness and developes
into an immortal entity.

With these lower forms there is no demur against reincarnation it is accepted blindly and instinctively and
without special conciousness of it. But with advancing
intelligence and power man becomes a creature of free

will and of moral responsibility, and he can as well choose whether or not to be reincarnated, as a denizen of earth who has the time, means and opportunity, can decide for himself whether he shall take a foreign trip or remain at home.

During the progress of the Christian era so called—the planetary spirits and ancients who had formerly peopled the earth decided that within the centuries of its second thousand of years, knowledge and light from the spirit world, in actual demonstration of individual immortality must be turned upon this world. And in their deliberations and plans as to the best method and place for this, attention was especially directed to America, for here the spirit of freedom had been imbibed by the various tribes and nationalities who had gathered here or who had lived and passed on, and because of the number of planetary and of Ancient spirits who had at one time or another either became reincarnated on its shores, or influenced individuals to come to the country from other lands.

And it was decided that as humanity was growing more and more sensitive and psychical on earth, and more susceptible to occult influences and forces that about the middle of the nineteenth century the demonstrations might be forcibly made through mortal media—of spiritual and tangible existence of intelligent decarnated beings, all of which was brought to pass as annals of modern spiritualism can prove.

Now be it known that through their deep interest in the dispensation of spiritual truth to man a number of the ancient spirits—not only of Atlantis but of other spheres and worlds decided to engage in the work of enlightment

and of general helpfulness to earth's people, some of these
were relegated to earth to become embodied as mortal entities to serve as educators and leaders to the people, as
soon as their mortal brains could be developed to receive
the inspirational force of their advanced guides or sufficiently unfolded in magnetic qualities to intelligently
generate and utilize the healing forces that could be imparted to the sick and suffering for healthful purposes.
Others of the ancients were to remain more or less in the
atmosphere of earth as guides to and co-workers with their
reincarnated friends, and others, again, were to bring
stimulating forces to various earth people from the
evolved kingdoms, who because of their ancestry and
inherited powers possessed qualities and elements that
could be drawn upon and utilized for the production of
various forms of startling psychical phenomena. And
thus many grades and qualities of spirit mediumship were
developed and the highest spiritual form and instructive
quality in the ethical line and grandest psychical perception including the highest type of prophecy—was of reincarnated media, and there were intelligent people, aspir-
ational and progressive, willing to be taught by the
higher school of ethical and thoughtful wisdom—and of
these were many inspired souls who were seers, prophets,
teachers, inspirational philosophers, musicians and clairvoyants, healers of the sick and humanitarian reformers,
of this class were Harma and Zaida, with many another
from their own sphere of culture and thought and as
teachers and guides were such as Iberna and Cabul and
others of the Atlantian race who claimed them as their
kin.

But the more degraded or ignorant of people—including media—who cared not to learn or to advance in knowledge or power in the higher spiritual cult, but who clung to the things of the past and to wordly show, were of the evolved class that had come up from the animal kingdoms and who yet patronized the instincts of the lower forms, yet under the guidance of high forces using intermediary spirits as agents in reaching these, types of mediumship were developed and utilized for the blessings of man.

XXXVI.

And through those types of mediumship even of the lower phases and exhibitions an evolutionary process and stimulating power runs, which produces year by year an unfoldment toward greater spiritual growth and strength of even the ignorant and—more or less—brutish class and of the most improved or ignorant of mediums.

And thus in the nineteenth century thousands of psychics were developed and utilized for the manifestation of decarnated intelligence to the world and a grand work for humanity which has stimulated to higher and nobler lines of thought and methods of labor has been wrought.

Now be it known that among the spirit entities who had taken an interest in the evolution of the race were thousands of those who had in bygone centuries belonged to the aboriginal tribes of North America—and the earlier tribes of these people had been kindly, simple, and affectionate, dignified in bearing and lofty in sentiment

and in the spirit world they had been educated and
trained for special good works.   Now as the atmosphere
of this America had become impregnated with the Indian
element and magnetism it was an easy matter for many
of the spirit Indians to work in it in association with
earthly psychics—and as this magnetic force of theirs
possessed vital healing qualities these spirit Indians could
freely impart it to the sick and cure them of their ills,
and also utilize it for the protection of their media in
various forms.

And thus hundreds of spirit Indians were selected as
guides and healers and messengers and sent out to the
most of the media of the country who as a rule were blessed and strengthened by their helpfulness, and spiritualism became a debtor to the red race.

But while many of these Indian guides or workers were
in fullness only Indians who had through the law of evolution, developed upward through various stages of physical and spiritual progress to their present places—many
of the more refined and intelligent were in reality original
planetary or Ancient beings of Atlantis or elsewhere who
had at one time been reincarnated for some special work
or influence among the most advanced tribes of North
American natives but who in the higher spheres are
known in their original form and spirit.   Yet these on
coming to earth life as guides and workers assume the
manner and appearance more or less cultivated in spiritual quality of their Indian incarnation.

These are usually to be found among the cultured Indian guides of advanced media who are healers and instructors of the more intelligent and spiritual class.

Millions of human beings on earth are reaching out in thought and aspiration to the unseen forces of the higher life, for guidance and knowledge. Many of these can only be satisfied with direct communication from their personal spirit friends and over and over again through the passage of years, they must receive the token of identity and the individual message from above in order to be led along any line of spiritual culture and thought. Others again, having once been assured of the living presence and intelligent communication of their blessed ones from on high, are not contented to repeat the experience over and over again but they branch out in new realms of thought and research into spiritual lore and law, and ask for lessons upon the philosophy of life and the mystery of being, from advanced minds in the spheres whom they deem competent to instruct and guide them—and just as there is a supply for every demand, so is there a response to the cry and need of these classes, and while the former receives from the seance the continued assurance of the love and nearness of their personal spirit friends—so the latter, sooner or later are elevated to a sphere of receptivity and vibration through which they are given the light and knowledge from ancient and modern teachers.

There are others of the ranks of progressive souls who can really find in spiritualism no source of supply for the demands of their natures. It's modern revelations are not mysterious and weird enough for their natures, and they are prone to clothe everything, that cannot be explained on physical grounds, and much that can, with the drapery of mystery and supernaturalism and there are very many manifestations in the universe that seem uncanny enough to satisfy even such wonder seeking souls·

And again there are those who love to dip into the realms of Ancient lore and cult, and they find in theosophy certain signs and revelations, theories, and symbols that satisfy their minds and they are content to dwell in the realm of mingled truth and of vagary as handed imperfectly down to humanity from the ancient lore of the Hindoos.

Still there are other souls seeking the light and in their aspirations for the high and good, they ignore the conditions and the demands of matter or of physical law and recognizing the truth that—in the light of the entire scheme of creation and unfoldment "all is good," they are prone to overlook the fact that in this developement and operation of the higher law, suffering is engendered for races and individuals that is not only very real, but which also must be treated in accordance with both the law of psychics and the law of mind. Yet all these classes are reaching out for the spiritual, and according to their seeking and their demand the responses come.

Ancient spirits of the higher spiritual spheres are busy working for the uplifting of the human race on earth, many of these intelligences belong to other planets yet they labor in conjunction with the people of earth and may continue to do so for still thousands of years to come.

The glory of this nineteenth century is not so much in the grand revealments of science as it is in the glorious demonstrations it has made of everlasting, individualized life for man. Standing out from its predecessors in modern history the nineteenth century flames with the

splendor of the spheres that has opened upon it the revelations of life and works in the world beyond, and Ancient
spirits are responsible for the work, for they, through
thousands of years have been instrumental in magnetizing
the inhabitants of earth with spiritual force that has
quickened the psychic nature of humanity, and helped to
bring it into rapport with the celestial spheres.

Tutelary Gods themselves plan great works, they oversee
planets and construct Zones, but it is their agents who
work out in detail much of the important labor that is
performed. Tutelary Gods are unseen by personages inferior to them at least by those who are not nearly on a
plane with themselves. Ancient Spirits can be seen and
recognized by many who have not reached their standard
or plane of exaltation and unfoldment, who receive of
their council and ministration, though the beings of the
lowest sphere in spirit do not behold them as personalities but rather as rays of light.

The work of planet building is a noble one, and as the
mortal child builds his toy house of cards or blocks, so
the Tutelary and his assistant frame a planet from the
material and the aura of worlds that he has at command.
But unlike the work of the infant which may be blown
away at almost a breath, the earth planet lives through
the ages, an enduring tribute of majesty and power, to
the intelligence, design and skill that gave it birth.

> Out of the shadow of the mighty past
>   The active brain and ceaseless hands of man
> Upon the wondrous Universe hath cast
>   The horoscope of God's almighty plan.

For Lo! the human hath developed well
   A selfhood that is evermore divine,
A potency no mortal tongue can tell—
   To wonders execute of great design,
Magnificent beyond all less degree
   The grand progressive power of the soul
And man exalted in his majesty
   Doth worlds and systems hold in his control.

He speaks, and lo, at his command
   The restless waves of force assume
The form and potency of being, planned
   By him whose splendors every form illume,
He wills, and lo the empty voids become
   The scene and centre of great moving orbs,
And all creation sings that once was dumb—
   And matter, spirit energy absorbs.

Lo! Man in moving upward calmly sweeps
   All dross and obstacle from out his path—
Eternal vigilance that never sleeps
   Affords him freedom e'er from Natures wrath,
All things become subservient to the mind
   That grasps the potency of spirit force,
And all things fleeting, here are left behind
   The soul majestic, in its onward course.

Planets and systems spring to active life
   'Neath the perfected power of human will
Man in potential force is ever rife,
   He can control the world through good or ill,

Man in completeness is a very God
In grandeur, ministry, and noble power
The pathway of the stars his soul hath trod
Infinity in essence is his dower!

XXXVII.

Every nation and every race that has ever existed upon the face of the earth have had their wise men and prophets, their advanced minds, their great moral and physical leaders who have led them on to heights of power or of knowledge, that the race or the Nation could not have attained without them. These foremost souls have exhibited a mentality, a prowess, and a spiritual cult, far in advance of the quality and progress of their times. They have been the standard bearers of truth, the heralds of progress, who have shown to the world, abilities and wisdom that could not have been caught or inherited from the ancestry from which they sprang upon the mortal plane.

In the direct line of evolution or of heredity.—Independent of the action of psychic law, these great minds could not have been born until the earth had passed through a thousand times as many ages as it has done. For the stream can not run higher than its source, an the line of human nativity—even through the elevating powers of evolution cannot possibly produce phenomenal expressions of an enlarged mentality that is beyond its natural power and scope. But the ages have given far advanced minds to the world. Minds that in th ir grand teachings and philosophical exegesis of the purpose and destinies of life have done much to quicken the higher principles of manhood and of spiritual cult in the hearts of humanity, which have assisted them to become more receptive to the influence and inspiration of the heavenly spheres.

Thus Budha, Confucius, Jesus, and countless other leaders and reformers in the ethical line of thought and culture, and hosts of leaders in other departments of human interest and human need have lived and labored for mankind in advance of the age to which they belong. They have done their work and passed into higher scenes of achievement and of power. Such souls have lived for ages in other bodies and amid other conditions. Such are the planetary workers of the spheres, or among the Ancients who dwelt upon old Atlantis or in even more remote times upon the earth. Such souls have been content to be reembodied as living entities in earthly form, for they realized the work demanded of them.

And such a history as theirs of repeated incarnations that they might bear an influence and a glory to earth and its people, independent of any power the planet could produce in its own unfolding life, is a history of magnificent expressions, of splendid achivements, and of holiest purposes for the love of humanity. Beside such a record, the desires and inclinations of human beings who deny the possibility of re-incarnation because it seemeth to their narrow view of life, to shut them out of a personal continuity of life or from their individal friends are as feeble and flickering as the rays of a glow worm compared to the glorious sun of Heaven short sighted mortals who cannot understand that the spirit *per se* never loses its identity. That forms and names are nothing more than to express identity for a season and that they change with every ascension of the spirit to a higher realm. Narrow minded mortals cannot comprehend that eternity is large enough and long enough for the ego to pass through any number of reincarnations on every planet that it desires to explore. Purblind creatures blinking at the light of the ages and shrinking from the dazling rays of truth, who cannot realize that soul kindred can never be separated, and that they who are really our own will continue to be so as long as the spiritual attractive force holds them to us and ourselves to them—and no longer.

And it may be as well for human beings to learn on earth some of the things that they must learn sometime, and one of these truths is that many who cling closely for a year or for a century or five centuries—for time counts but little beyond—may, in time, outgrow the attractive force and psychical association that binds them together and go into other spheres where entirely new associations and social ties are formed. But whatever is of the soul lives and remains a part of our life—on earth friends sometimes grow apart, even kindred, not because of discord or difference of opinion, but because they have each absorbed or given forth to the other, all the magnetic force or psychical elements that can be thus utilized and attractive force that drew them together has lessened its power and ceased to operate as a magnet. And in this there is no ground for complaint nor quality of discontent. It is only the operation of natural law. Such human beings who have ceased to cling to each other in the closest ties of affection or of friendship may always wish one another well, and each be interested in the welfare of the other, but each can only hold to the ties and relationships that the soul unfoldment prepares and fits them for, and therefore it is by no means singular that many of your friends and neighbors whom you once knew and admired—or vice versa have drifted away, either on earth or on the planes of spirit life and that no communication or correspondence passes between you.

The Ancients understand the operations of all these laws and they work along lines of the least resistance to them. Therefore they are in harmony with natural law, they are seeking for the greatest good for the largest number and they are above all petty personalities or fancies, having great things to do to ocupy their attention and time.

Great souls are not those who live in the selfishness of personal gratification, who are looking forward to the culmination of hopes and plans and ends by which they may attain some individual happiness or aim. They are those who bend energies and talents in the work and direction of human interest for the struggling masses who are plowed under by the implements of suffering and despair. Yet great souls do not wholly ignore the personal comfort and needs of individuals, but these are made secondary to the elevation and unfoldment of the many.

One cannot however expect a child of earth who has lived but a few years more or less amid contending conditions to look upon life through the philosophical glasses of the Sages and Prophets of the mighty past who are ages wise in experience growth and achievement. To these Ancients the most advanced mind on earth—the most intellectual brain and highly cultured nature, must seem crud? and simple compared to their own quickened mentality and elegant culture, while the man of four score years can only be counted as a mere babe in the scale of experience that holds the accumulated wisdom and knowledge of the patriarch of ten thousand years. Oh wonderful, wonderful soul-germ that in the original start of its course as a living entity bears with it from the great source of all life and light, the potency of a never failing energy and use.

Oh wonderful soul-germ that by the law of attractive force pursues its way into the very midst of elemental forces and conditions aggregating to its self the necessary essentials for the upbuilding of an intellect and the creation of a useful form! Oh wonderful soul-germ, that sweeps in to the midst of atoms and ethers, bearing all before it with resistless power until it animates a breathing mortal form and quickens it with the breath of eternal life.

"What is man—Oh God!—that thou should be mindful of him", Yea, oh mortal, Soul—ask it of the winds that bear your questionings to the uttermost parts of the earth. Ask it of the waves that rush in ceaseless force from shore to shore. Ask it of the stars that roll and burn and sing their songs of everlasting praise to Life, eternal Life. Ask it of the Tutelary Gods who guide worlds and fasion planets by the skill of created thought, electrical force and of vitalized energy. Ask it of planetary beings who move in majesty with revolving worlds·

Ask it of the Ancients who watch the ages roll away and whose wisdom surpasses the power of voice to repeat. Ask it of the Supernal whose body the Universe is—What is man ? And the responses come from every questioned point and force—Man is divine! Man is power! Man is personified Deity! Man is everything! and God may well be mindful of him and guide him on to perfect glory and unfoldment—for man is a part and the life of God.

XXXVIII.

Poor shortsighted mortals who hold fast to the things and customs of the past, or even ye who cling to the affairs and the practice of today, loth to change with the progressive advancements of the times—ye know not what ye do. Thousands upon thousands of human entities like yourselves are clinging to the rock of ages, or standing ankle deep so to speak—in the clay of old time habits, thoughts, and prejudices, instead of soaring as immortal spirits through the everlasting spheres. Nor are these thousands encased in earthly forms for they have centuries or decades—it matters not which—ago cast off the physical form and stepped into the spirit world. Yet they are fossils holding to the old, petrified in the ruts of self conceit and opinion, unmindful of the grander light and larger scope of the spiritual towards which they may soar in beauty and with energy, if they but try their powers and attempt to swing off towards the boundless spheres with the glory of God in view.

But the timid, clinging entities who perhaps are for

the first time travelling over the pilgrimage of the earthly life may be pardoned if they do not see in the stars and planets that roll above them the worlds that they may in time inhabit if they desire and work for it, or the zones and orbs of living glory that such beings as they, in the perfection of power, can create. Therefore are they objects of compassion and of instruction of the Ancient characters who have for thousands of years been travelling the chief highways of the stars and practising with the forces and elements of worlds for the achivement of grand things. So are hosts of earth bound spirits who cannot or will not soar, creatures of pity and ministration for the higher intelligences to attend. Thousands who cling to the old life, to the old attractions that belong only to the material to old customs and errors are the object of instruction of these Ancients. But it is not all at once that they can be moved from their ruts, or vitalized by the magnetism of the higher, into new action and unfoldment, and scores of ancient spirits are working in harmony with other helpers and guides of the more modern day to bring these hosts of enslaved beings to a higher comprehesion and a loftier state of thought and progress.

Many mortals who have learned of spirit communication have wondered why ancient spirits should be attracted back to earth and why they should wish to deal with the affairs of men, and the question is repeatedly asked. Why these ancients do not pass so far beyond earth that they can have no power of attraction back to its environments.

Now be it known that only those of the ancients who

were powerful leaders in the sciences of mortal philosophy, of metaphysics of physical law and its operations,
who were in the past, teachers, seers, and advanced
minds do return to earth and guide and direct their forces
in their unfoldment and to bless human beings individually and collectively by their uplifting influence and their
instructive force.   Many others who have advanced in
great spiritual exaltation and wisdom spend their time
and energy in helping and teaching spirits who have passed from the earth body but who dwell in the three or
four lower spheres of spirit life.   While others of the Ancients have passed on and on from Zone to Zone and are
filling their places and doing their part in the great scheme
of being, never returning to earth or giving any special
attention to its inhabitants.

The ancient intelligences who have direct access to
earth and who have communication with or influence over
certain of its people, are in direct spiritual relationship
with kindred souls who are reincarnated on this planet.
Ties of true consanguinity bind them together.   Therefore the attractive force for the decarnated Ancients who
come to earth is a most natural one, for love follows its
own from Zone to Zone, and binds the souls of those who
are of one sphere or kin together.

The Ancients who come to earth are those also who
have breathed their magnetic forces upon the planet;
the very atmosphere is impregnated with their electrical
force, the sunshine is tinted with the magnetism of their
personality, and the soul of the globe has caught its radiant energy from them and the planetary beings who have
assisted in its unfoldment and stimulated its evolving
power.

They are the entities who labor long and faithfully for the uplifting of a world and for the perfecting of a humanity.

Races come and go, nations rise and fall, dynasties ascend to power and are swept away like pillars of sand in a bracing wind, but these human entities work calmly on undisturbed by storm or shine, waiting the time when man in awakened power and perfected selfhood shall dwell upon the planet in all the pristine glory and vigor that the planetary beings who first took upon themselves forms of matter and came to earth to vitalize it with their own potential force, possessed.

There are thousands of media on earth who are sensitives, media who have lived in other forms in bygone ages, media who are here to do an humble or a mighty work. Each fits into the place assigned for him or her. Each is working out an important part in personal destiny and at the same time doing the bidding of the higher law and force. All media are not reincarnated souls many of them have lived before as conscious working entities. The fact is kept from many of them concerning their preexistence for the work and the duty of the present is all that they need to know. Each of the re-embodied media is attended by an Ancient Spirit, who watches over and helps to direct and to shape affairs that will bring the sensitive certain needful and valuable experiences. Every child of earth—humble or of high degree—has spirit guides and watchers. All do not have or attract either ancient or planetary attendants. The guides of many mortals are friends from their own associates and relatives with one or two more advanced teachers of a

modern age, and in guarding their earth friends and seeking to exercise wise judgment over them with forethought of their welfare, this class of guides is winning
for itself a developement and an experience that is fitting
them for the higher—and still higher—realms.

It is through labor, through unselfish thought and
action, through growth and aspiration that souls arise.

What the most glorious and advanced ancient has become may be reached by spirits who still walk the pathways of earth.

What the Arch Angels and Seraphs of the supernal
spheres are in light, intelligence and power of achivement
any one of earth who reads these lines may aspire to and
reach.

The path of progress is forever on, and human entities
who enter upon it may find unnumbered opportunities
opening before them to thrive, to grow, to learn, to unfold, to labor and to achieve. There is a vast realm of
possibility still opening before the most progressive soul
—a realm of discovery and of power. A realm of un-
explored beauty and of grandeur, and the limitless freedom
of eternity affords to the ego that scope and force that he
needs to spur him on and on to reach perfection.

Sometimes it is asked of us if the sun of your solar system is inhabited. The sun is not only a great luminary
but it is also a great reservoir of electrical force which
sends out its currents with tremendous energy and force.
Human beings cannot live within its atmosphere.

The radiant energy is that which the sun supplies to its
system in the form of force and motor power. The
electrical fluid is absorbed in part by tutelary and plane-

tary spirits but they do not dwell upon the sun. Yet we are told that the sun is made up in part of the essence of spiritual force that is of itself the life of soul-germs.

This essence generates an aura that enters into the composition of the spiritual sun the counterpart of the more physical luminary, and this is vitalized by that essence so that soul-germs find upon it an atmosphere which they can dwell within until they are drawn by the irresistible force of attraction to earth or some other planet, there to vitalize elements of matter and of ether and to correllate them together until the union of these with soul-force become eternal spirit.

Every part and portion of the universe has its place and use—every atom that has ever existed has filled a part in the great scheme of life. Every pulsation of being every vibration of force has its mission and its work. All is divinely planned and wisely executed.

The law of evolution is a noble law, it developes even the highest forms of existence and of intelligence from the lowest strata and condition of being. Causation is the supreme power that governs all things — effect in every department of life and action is but the result of causation. Every entity however lowly is an heir to the highest attainment though it take a million years to reach the throne of wisdom and of light.

For even the Tutelaries, the planetary beings, the ancient of ancient souls have all been evolved from primal soul germs and simple forms of physical beings upon some planetary Zone.

"The universe is one stupendous Whole
Whose body, Nature is, and God the Soul."

www.ingramcontent.com/pod-product-compliance
Lightning Source LLC
Chambersburg PA
CBHW030126030726
47498CB00007B/2577